DATE DUE

DEVELOPMENTS IN SEDIMENTOLOGY 15

THE CHEMISTRY OF CLAY MINERALS

FURTHER TITLES IN THIS SERIES

1. *L.M.J.U. VAN STRAATEN, Editor*
DELTAIC AND SHALLOW MARINE DEPOSITS

2. *G.C. AMSTUTZ, Editor*
SEDIMENTOLOGY AND ORE GENESIS

3. *A.H. BOUMA* and *A. BROUWER, Editors*
TURBIDITES

4. *F.G. TICKELL*
THE TECHNIQUES OF SEDIMENTARY MINERALOGY

5. *J.C. INGLE Jr.*
THE MOVEMENT OF BEACH SAND

6. *L. VAN DER PLAS Jr.*
THE IDENTIFICATION OF DETRITAL FELDSPARS

7. *S. DZULYNSKI* and *E.K. WALTON*
SEDIMENTARY FEATURES OF FLYSCH AND GREYWACKES

8. *G. LARSEN* and *G.V. CHILINGAR, Editors*
DIAGENESIS IN SEDIMENTS

9. *G.V. CHILINGAR, H.J. BISSELL* and *R.W. FAIRBRIDGE, Editors*
CARBONATE ROCKS

10. *P. McL. D. DUFF, A. HALLAM* and *E.K. WALTON*
CYCLIC SEDIMENTATION

11. *C.C. REEVES Jr.*
INTRODUCTION TO PALEOLIMNOLOGY

12. *R.G.C. BATHURST*
CARBONATE SEDIMENTS AND THEIR DIAGENESIS

13. *A.A. MANTEN*
SILURIAN REEFS OF GOTLAND

14. *K.W. GLENNIE*
DESERT SEDIMENTARY ENVIRONMENTS

DEVELOPMENTS IN SEDIMENTOLOGY 15

THE CHEMISTRY OF CLAY MINERALS

BY

CHARLES E. WEAVER

School of Geophysical Sciences, Georgia Institute of Technology, Atlanta, Ga. (U.S.A.)

AND

LIN D. POLLARD

Emory University, Atlanta, Ga. (U.S.A.)

ELSEVIER SCIENTIFIC PUBLISHING COMPANY
Amsterdam — Oxford — New York, 1975

ELSEVIER SCIENTIFIC PUBLISHING COMPANY
335 JAN VAN GALENSTRAAT, P.O. BOX 211, AMSTERDAM, THE NETHERLANDS

AMERICAN ELSEVIER PUBLISHING COMPANY, INC.
52 VANDERBILT AVENUE, NEW YORK, NEW YORK 10017

FIRST EDITION 1973
FIRST REPRINT 1975

LIBRARY OF CONGRESS CARD NUMBER: 72-83214
ISBN 0-444-41043-0

WITH 32 ILLUSTRATIONS AND 75 TABLES

COPYRIGHT © 1973 BY ELSEVIER SCIENTIFIC PUBLISHING COMPANY, AMSTERDAM

ALL RIGHTS RESERVED. NO PART OF THIS PUBLICATION MAY BE REPRODUCED, STORED IN A RETRIEVAL SYSTEM, OR TRANSMITTED IN ANY FORM OR BY ANY MEANS, ELECTRONIC, MECHANICAL, PHOTOCOPYING, RECORDING, OR OTHERWISE, WITHOUT THE PRIOR WRITTEN PERMISSION OF THE PUBLISHER, ELSEVIER SCIENTIFIC PUBLISHING COMPANY, JAN VAN GALENSTRAAT 335, AMSTERDAM

PRINTED IN THE NETHERLANDS

PREFACE

In 1962 Dr. Earl Ingerson of the United States Geological Survey asked the senior author to write a review of the chemistry of the clay minerals to include in a new *Data of Geochemistry*. As there was no existing detailed review of the clay minerals, I decided to try to make the review critical and fairly comprehensive. A draft version was sent to Dr. Michael Fleischer (new editor of *Data of Geochemistry*) in 1968.

Dr. Fleischer was concerned about the length of the review but agreed to publish it as a Professional Paper; however, he recommended that I try to get it published by a commercial publisher. In 1970, Dr. Lin Pollard kindly agreed to help me up-date and finalize the original draft. This work was completed at the end of 1971.

The book is not as comprehensive as I would like. In part, this is because the original request was to make a selected review and not stray far from chemistry. I think the book will serve a purpose and will be a base that will make future reviews easier.

Thanks are extended to Wes Rice of the Continental Oil Company for assisting with the computer programming, Horace Bledsoe, Jr. for doing the drafting and Dianne Clark for doing the typing. The assistance of John Hathaway, Earl Ingerson and Mike Fleischer is also deeply appreciated. A number of people volunteered original data which is acknowledged in the text.

The senior author dedicates this book to Janice, Alaine, Patrice and Allison for their forbearance.

C.E.W.

CONTENTS

PREFACE V

CHAPTER 1. INTRODUCTION 1

CHAPTER 2. ILLITE 5
Illite origin 19
Sericite 21

CHAPTER 3. GLAUCONITE 25
Composition 25
Age 39
Origin 43
Non-marine glauconites 44

CHAPTER 4. CELADONITE 47

CHAPTER 5. SMECTITE 55
Dioctahedral smectite 55
    Montmorillonite 55
    Nontronite 75
Trioctahedral smectite 77

CHAPTER 6. CHLORITE 87
Macroscopic chlorite 87
Trioctahedral clay chlorite 91
Dioctahedral clay chlorite 94

CHAPTER 7. VERMICULITE 99
Macroscopic vermiculite 99
Clay vermiculite 102

CHAPTER 8. MIXED-LAYER CLAY MINERALS 107
Illite-montmorillonite 107
Chlorite-montmorillonite 114
Other mixed-layer clay minerals 118

CHAPTER 9. ATTAPULGITE AND PALYGORSKITE 119

CHAPTER 10. SEPIOLITE 127

CHAPTER 11. KAOLINITE 131

CHAPTER 12. DICKITE AND NACRITE 145

CHAPTER 13. HALLOYSITE 149

CHAPTER 14. ALLOPHANE 155

CHAPTER 15. TRIOCTAHEDRAL 1:1 CLAY MINERALS 159

CHAPTER 16. LOW-TEMPERATURE SYNTHESIS 169

CHAPTER 17. RELATIONS OF COMPOSITION TO STRUCTURE 173
Trioctahedral sheet 173
Dioctahedral sheet 175
Dioctahedral and trioctahedral sheets 176
Octahedral–tetrahedral relations 178
Chain structure 186

REFERENCES 189

INDEX 207

Chapter 1

INTRODUCTION

Clay minerals occur in all types of sediments and sedimentary rocks and are a common constituent of hydrothermal deposits. They are the most abundant minerals in sedimentary rocks perhaps comprising as much as 40% of the minerals in these rocks. Half or more of the clay minerals in the earth's crust are illites, followed, in order of relative abundance, by montmorillonite and mixed-layer illite-montmorillonite, chlorite and mixed-layer chlorite-montmorillonite, kaolinite and septachlorite, attapulgite and sepiolite. The clay minerals are fine-grained. They are built up of tetrahedrally (Si, Al, Fe^{3+}) and octahedrally (Al, Fe^{3+}, Fe^{2+}, Mg) coordinated cations organized to form either sheets or chains. All are hydrous.

The basic structural units in layer silicates are silica sheets and brucite or gibbsite sheets. The former consist of SiO_4^{2-} tetrahedra connected at three corners in the same plane forming a hexagonal network. The tips of the tetrahedra all point in the same direction. This unit is called the tetrahedral sheet. The brucite or gibbsite sheet consists of two planes of hydroxyl ions between which lies a plane of magnesium or aluminum ions which is octahedrally coordinated by the hydroxyls. This unit is known as the octahedral sheet. These sheets are combined so that the oxygens at the tips of the tetrahedra project into a plane of hydroxyls in the octahedral sheet and replace two-thirds of the hydroxyls. This combination of sheets forms a layer.

The major subdivision of the layer lattice silicates is based upon the type of combinations of the tetrahedral and octahedral sheets. Additional subdivision is based on:

(1) whether the octahedral sheet contains two cations per half unit cell (dioctahedral) as in gibbsite or three cations per half unit cell (trioctahedral) as in brucite;

(2) the manner of stacking of the tetrahedral–octahedral units upon each other;

(3) the amount and type of isomorphous replacement of the cations.

The 1:1 clay-mineral type consists of one tetrahedral sheet and one octahedral sheet. These two sheets are approximately 7 Å thick. This two-sheet type is divided into kaolinite (dioctahedral) and serpentine (trioctahedral) groups. The kaolinite minerals are all pure hydrous aluminum silicates. The different members are characterized by the manner of stacking of the basic 7 Å layers (Brindley, 1961b).

The trioctahedral two-sheet minerals are called serpentines. The serpentine minerals (chrysotile and antigorite are the most common), which are included in this subgroup, consist of a tetrahedral sheet and an octahedral sheet containing magnesium with only minor amounts of aluminum. The other minerals in this subgroup have a wide range of variations in composition. Aluminum, iron, manganese, nickel and chromium can substitute for magnesium in the octahedral sheet and aluminum, ferric iron, and germanium for silicon in the tetrahedral sheet (Roy and Roy, 1954). The serpentines in normal sediments usually occur mixed with kaolinite and/or chlorite and are difficult to identify.

The three-sheet or 2:1 layer lattice silicates consist of two silica tetrahedral sheets between which is an octahedral sheet. These three sheets form a layer approximately 10 Å thick. The oxygens at the tips of the tetrahedra point towards the center octahedral sheet and substitute for two-thirds of the octahedrally coordinated hydroxyls. The 2:1 clay minerals include the mica and smectite groups which are by far the most abundant of the clay minerals. The pure end members of this type are talc, a hydrous magnesium silicate; pyrophyllite, a hydrous aluminum silicate; and minnesotaite, a hydrous iron silicate.

The 2:1 structural unit of the micas is similar to that for talc; however, between the 2:1 units is a plane of large cations. These are referred to as interlayer cations. Potassium is the most common, but sodium and calcium also occur. These interlayer cations fit into the hexagonal ring formed by the tetrahedral oxygen ions and bond adjacent 2:1 units. The interlayer cations balance the charge due to the substitution of cations of lesser charge for some of those of greater charge in the tetrahedral and/or octahedral sheet. The basic 2:1 units are stacked together in a variety of stacking sequences (polytypes) : 2M (two-layer monoclinic), 1M (one-layer monoclinic), 1Md (disordered one-layer monoclinic), and 3T (three-layer trigonal). The first three are by far the most common. Muscovite is commonly the 2M type, phlogopite the 1M, and mixed-layer clays the 1Md.

The mica group is subdivided on the basis of whether the species are dioctahedral (muscovite type) or trioctahedral (biotite type). The micas are further characterized by the number of silicon ions in the tetrahedral position: tetrasilicic, trisilicic, disilicic, and monosilicic. Aluminum and less commonly ferric iron substitute for the silicon. The micas are further categorized according to the wide range of cations and the combination of cations which occur in the octahedral sheet. Aluminum (muscovite) and magnesium (phlogopite) are the only two which occur alone in the octahedral sheets. Most micas have two or more cations in the octahedral sheets; aluminum, magnesium and iron occur in a variety of combinations in the octahedral sheet. Mn, V, Cr, Li, Ti and a variety of other cations can occur in varying amounts. The substitution of a cation of a lower charge for a cation of a higher charge in both the octahedral (e.g., Mg^{2+} replacing Al^{3+}) and tetrahedral (e.g., Al^{3+} replacing Si^{4+}) sheets, gives the 2:1 layer a net negative charge which is satisfied by the interlayer cations. There is a long list of mica names (Foster, 1956; Warshaw and Roy, 1961) which have been established from the study of coarse-grained minerals. Most of these minerals probably exist in argillaceous sediments but their identification is extremely difficult.

INTRODUCTION

The fine-grained micas belong to the illite family. The dioctahedral illites greatly predominate over the trioctahedral. The most common illite mineral is dioctahedral and has approximately half as much aluminum substituting for silicon in the tetrahedral sheet as does muscovite. Approximately three-fourths of the octahedral cations are aluminum; minor amounts of ferric iron are present, and approximately one-eighth of the cations are divalent (magnesium and ferrous iron). This gives a total negative charge of approx. 0.75 as compared with a value of 1 for muscovite (per $O_{10}(OH)_2$).

The dioctahedral iron illites are the minerals glauconite and celadonite. Glauconite is also used as a rock name and is applied to any aggregate of fine-grained, green, layer minerals. Similar to the aluminum illites the iron illite layers commonly occur interlayered with montmorillonite-like layers. In glauconites more than half the octahedral positions are filled with iron, the more abundant being ferric iron. The aluminum content of the tetrahedral sheet is usually less than that of the aluminum illites and the magnesium content of the octahedral sheet greater. Celadonite has more octahedral Mg and less tetrahedral Al than glauconite. Trioctahedral illites are relatively rare as pure minerals concentrates but have been reported in Scottish soils (Walker, 1950). Because the interlayer cations are relatively easily leached from biotite, the clay-sized minerals are usually mixed-layer biotite-vermiculite (expanded biotite).

The expanded or expandable 2:1 clay minerals vary widely in chemical composition and in layer charge. These minerals are characterized by the presence of loosely bound cations and layers of water or polar organic molecules between the silica sheets. The interlayer width is reversibly variable. The interlayer water can be driven off at temperatures between 120° and 200°C. Sodium, calcium, hydrogen, magnesium, iron, and aluminum are the most common naturally occurring interlayer cations.

The dioctahedral subgroup is by far the most abundant. The layer charge on the expanded clays ranges from 0.3 to 0.8 per $O_{10}(OH)_2$ unit of structure. The low-charged (0.3–0.6), expanded minerals are called montmorillonite, montmorillonids, and smectites, among others. Subdivision of the expanded clay group is still a problem.

The low-charge (0.3–0.6) dioctahedral minerals which have most of their charge originating in the octahedral sheet are called dioctahedral smectites or montmorillonites. Expanded dioctahedral smectites, which have a relatively high tetrahedral charge content (0.4), are called beidellites. These clays commonly have a total charge of 0.7 or higher. Identical minerals, which are known to have been derived by leaching potassium from illite or muscovite, are referred to as dioctahedral vermiculite. There is apparently a complete gradation in composition and charge between the species of montmorillonite and beidellite. The ferric iron-rich variety is called nontronite.

Although they are rare in sediments, there is a wide spectrum of trioctahedral expanded clays. The most common in the low-charge range (0.3–0.5) are hectorite, which contains magnesium and lithium in the octahedral sheet, and saponite, which has considerable magnesium in the octahedral sheet and some aluminum substitution in the tetrahedral sheet.

Trioctahedral expanded 2:1 minerals with a layer charge of 0.6–0.8 are called vermiculites. These minerals are usually coarser grained and have better crystal organization than most expanded clays. The decision as to whether a clay should be called a vermiculite or not is usually based on its ability to adsorb two layers of ethylene glycol and expand to 17 Å. Walker (1958) has shown that magnesium-based expanded clays with a charge greater than 0.6 will only adsorb one layer of glycerol and expand to 14.3 Å (vermiculite); those with a smaller charge will adsorb two layers and obtain a 17 Å thickness (trioctahedral smectite). The present classification of the expanded clays and the illites is not at all satisfactory.

All the non-expanded 2:1 and 2:1:1 layer silicates can have their interlayer cations removed. Water and organic molecules can then penetrate between these layers to form expanded layer minerals.

Chlorite can occur as a clay-sized mineral. Most consist of a 2:1 talc layer plus a brucite sheet. This forms a unit 14 Å thick. Most chlorites are trioctahedral although a few dioctahedral chlorites have been found. Some chlorites have both dioctahedral and trioctahedral sheets. Because substitution can occur both in the 2:1 layers and in the brucite sheet, the chlorites have a wide range of compositions. The coarser grained chlorites have been analyzed and classified (Hey, 1954) but relatively little is known of the composition of sedimentary chlorites.

There is a large number of clays which are not pure mineral types but consist of interstratified units of different chemical composition. (In detail, this may include nearly all the 2:1 layer minerals.) These are called mixed-layer clays. The two or possibly three different units can be regularly interstratified ABABAB or more commonly randomly interstratified AABABBABA. The most common regularly interstratified clay mineral, corrensite (Lippman, 1954), consists of alternate layers of chlorite and vermiculite or chlorite and montmorillonite.

Mixed-layer illite-montmorillonite is by far the most abundant (in the vicinity of 90%) mixed-layer clay. The two layers occur in all possible proportions from 9:1 to 1:9. Many of those with a 9:1 or even 8:2 ratio are called illites or glauconites (according to Hower, 1961, all glauconites have some interlayered montmorillonite) and those which have ratios of 1:9 and 2:8 are usually called montmorillonite. This practice is not desirable and is definitely misleading. Other random mixed-layer clays are chlorite-montmorillonite, biotite-vermiculite, chlorite-vermiculite, illite-chlorite-montmorillonite, talc-saponite, and serpentine-chlorite. Most commonly one of the layers is the expanded type and the other is non-expanded.

Attapulgite and sepiolite are clay minerals with a chain structure. The former has five octahedral positions and the latter either eight or nine. Both have relatively little tetrahedral substitution. The octahedral positions in sepiolite are filled largely with Mg and those in attapulgite with approximately half Mg and half Al.

Chapter 2

ILLITE

Pure monomineralic clay mineral samples are difficult to find. Even in instances where a sedimentary formation contains only one clay type, the presence of non-clay minerals makes purification difficult. Actually, it has not been established, particularly in the 2:1 and 2:2 (chlorite) clay groups, that the chemical compositions of clay minerals in sedimentary deposits are consistent over intervals of a few feet or even a few inches. Most clay minerals probably have considerable variation in composition from one unit cell to the next. It has been established that in many clays the strength with which interlayer cations are held varies widely from layer to layer. In other clays, such as the kaolins, isomorphous substitution may occur in only one unit cell out of ten or even one out of a hundred. Thus, any chemical analysis of a clay mineral is only an average and there is little chance of determining the range and standard deviation until analyses can be made at the unit-cell level. Considering the nature of the origin of most clay deposits it might be expected that the composition of sedimentary clay minerals would show considerable variations over relatively short distances.

Table I contains seven chemical analyses of "purified" Fithian illite (type illite—Grim et al., 1937). The variations are large as it was not possible to exercise any control over these analyses. The variation is a composite of all the errors inherent in any analytic analysis: operator, instrument, method, beneficiation, sampling. This is in addition to real differences. Unfortunately, this statement can be made about most clay-mineral analyses. There are relatively few duplicate analyses in the literature and little effort has been made to determine sub-sample variation.

Table II lists 29 illite analyses. The statistical data (Table III) were calculated using only the first 24 analyses; the others were added after the calculations were completed. The Fithian illite analyses show larger standard deviations for SiO_2, Al_2O_3, FeO, CaO, TiO_2, and H_2O than the 24 illites; the remaining values are larger for the 24 illites. The list includes those oxides most likely to be present as impurities so it might be expected that the sub-sample variation would be larger. The other elements, except for Fe_2O_3, are less likely to be present as contaminants and the large standard deviation suggests that these elements will be most useful in establishing differences within the illite group.

The samples range in age from Recent to Precambrian. Both pure 1M (No.27), 2M (No.6) and mixtures of the two polytypes along with varying amounts of 1Md are

TABLE I

Chemical analyses and statistical data on Fithian illite

	1	2	3	4	5	6	7	Average	Standard[1] deviation	Relative[2] deviation	Range	Maximum difference	Max. relative[3] deviation
SiO_2	51.22	56.91	48.10	45.7	51.4	56.1	52.75	51.74	4.01	7.75	56.91–45.7	11.21	21.6
Al_2O_3	25.91	18.50	24.61	30.8	24.2	19.0	24.83	23.98	4.20	17.51	30.80–18.50	12.30	51.2
Fe_2O_3	4.59	4.99	4.60	7.2	6.7	5.5	4.12	4.57*	0.35	7.66	4.99– 4.12	0.87	19.0
FeO	1.70	0.26	2.15	—	—	—	0.26	1.09*	0.78	71.56	2.15– 0.26	1.89	173.4
MgO	2.84	2.07	2.40	1.3	1.4	1.6	2.29	1.99	0.48	24.12	2.84– 1.3	1.54	77.4
CaO	0.16	1.59	0.20	0.6	1.8	2.1	0.32	0.97	0.83	85.57	2.10– 0.16	1.94	200.0
Na_2O	0.17	0.43	0.34	0.3	0.4	0.56	0.35	0.36	0.13	36.11	0.56– 0.17	0.39	108.3
K_2O	6.09	5.10	6.13	6.0	5.5	4.6	5.71	5.59	0.57	10.20	6.13– 4.6	1.23	22.0
TiO_2	0.53	0.81	0.50	—	—	0.93	0.62	0.68	0.41	60.29	0.93– 0.53	0.40	60.3
H_2O^+	7.14	5.98	6.62	8.1	7.9	7.3	7.94	6.92*	0.83	11.99	7.94– 5.98	1.96	28.3
H_2O^-	1.45	2.86	4.42	—	—	—	0.85	2.40*	1.53	63.75	4.42– 0.85	3.57	148.8
Total	100.70	99.50	100.07	100.0	99.3	97.69	100.04	99.76					

1. Grim el al. (1937).
2. Kerr et al. (1950).
3. Brindley and Udagawa (1960).
4. Prof. Taro Takahashi, Alfred U. (Wards Nat. Sci. Estab. sample.)
5. Prof. Taro Takahasi, Alfred U. (Am. Petroleum Inst, Ref. Clay No.35.)
6. Brannock (1960).
7. Whitehouse and McCarter (1958).

[1] Standard deviation = $\sqrt{\Sigma d^2/(n-1)}$; [2] relative deviation = (100 × standard deviation)/average;
[3] max. relative deviation = (100 × maximum difference)/average; *average includes only 4 analyses.

present and the Balleter illite (No.11) has a 3T structure; they have been collected from soils, marine muds, limestones, sandstones, and shales. It is difficult to show any relation between composition and any of these factors because of the small number of analyses, the limited amount of geological and mineralogical information, and the fact that most illites in sediments are at least second-cycle. However, there is a decided tendency for MgO to be more abundant in "marine" illites than in those formed under non-marine conditions (Table IV). The MgO content of the "marine" illites is similar to that of montmorillonites and may indicate these illites formed from montmorillonite, probably during burial. The low-MgO illites are primarily formed from the alteration of feldspar, as might be expected. Thus, Mg may be used to give information on the origin of illites. The low-MgO illites all have more than 1.70 octahedral positions filled with Al (Table V). This is within the range of low-Al muscovites.

Structural formulas were calculated using the method of Ross and Hendricks (1945). Table V gives the structural formulas of the 29 illites and Table VI contains the mean and other statistical parameters of the first 24 illites. The structural formula based on an average of the seven Fithian illite analyses is also included. Table VII lists the range of values for the Fithian illite and for the other 29 illites. For most structural positions the spread is larger for the 29 illites than for the 7 Fithian illites; however, tetrahedral Al, perhaps the most significant factor, has a larger range of values for the 7 than for the 29.

The average number of ions in octahedral coordination (Σ) is 2.07 with values ranging from 1.85 to 2.24. Much of the variation from the theoretical value of 2.00 is due to the presence of iron, aluminum, and magnesium external to the layer. Some of this material is present in chloritic interlayers.

The average of the calculated layer charges for the 29 illite samples is 0.72 per $O_{10}(OH)_2$ with values ranging from 0.43 to 1.20. These extreme values, as well as many of the other values, are probably incorrect. In all but one sample (No.21) the calculated layer charge and the sum of the external cations check quite closely. This does not mean that either value is correct. It is quite likely that many of the octahedral cations in excess of 2.00 per unit half cell occur as in the interlayer position. If interlayer cations are assigned to the octahedral sheet and are in excess of the required 2.00 cations per unit half cell, the charge balance will not be upset.

If the divalent cations in excess of 2.00 in the octahedral sheet are assigned to the interlayer position, the total layer charge increases as does the total number of interlayer cations (Table VIII). The values agree quite closely and the total layer charge values range from 0.81 to 0.94 with an average near 0.9.

Although a structural formula derived from averaging 24 structural formulas is given in Table VI, this is probably not a typical formula for a well-crystallized 2M illite. The illite from the Belt (No.8) has extremely sharp X-ray reflections and affords an excellent 2M pattern. This sample is probably one of the best crystallized and purest 2M illites described in the literature and the structural formula is close to being

TABLE II

Chemical analyses of some illites

	1	2	3	4	5	6	7	8	9	10	11	12	13	14
SiO_2	51.22	50.72	49.21	49.21	51.64	49.67	51.50	50.55	49.10	50.10	51.26	47.55	47.76	48.2
Al_2O_3	25.91	25.84	28.91	28.88	25.29	27.31	21.40	26.14	25.00	25.80	30.15	32.45	26.13	24.6
Fe_2O_3	4.59	4.57	2.31	2.34	4.32	2.96	1.60	0.67	7.50	2.97	2.36	0.76	5.66	7.2
FeO	1.70	1.21	0.02	0.01	0.72	0.00	0.00	0.65	0.00	0.00	0.59	1.85	0.00	1.5
MgO	2.84	2.65	3.33	3.32	2.90	1.09	3.50	4.25	2.10	2.95	1.37	1.70	3.56	3.
CaO	0.16	0.15	0.24	0.24	0.11	0.29	1.20	0.60	1.10	0.39	0.00	0.06	0.00	0.0
Na_2O	0.17	0.17	0.16	0.15	0.19	0.10	0.20	0.19	0.20	0.18	0.13	1.05	0.53	0.1
K_2O	6.09	6.14	7.21	7.20	6.31	7.26	11.00	10.29	7.30	7.99	7.77	6.22	6.81	5.9
TiO_2	0.53	0.45	0.51	0.50	0.44	0.23	0.22	0.42	0.90	0.94	0.15	0.64	0.00	0.0
H_2O^+	7.14	8.28	8.02	8.07	7.70	9.00	7.44	4.59	4.90	5.33	6.28	7.73	8.92	7.4
H_2O^-	1.45	0.21	1.22	1.16	0.30	1.50	1.83	0.99	1.10	1.93	0.00	0.00	0.63	1.6
Total	101.80	100.39	101.14	101.08	99.92	99.41	99.89	99.34	99.20	98.58	100.06	100.01	100.00	99.0

1. Grim et al. (1937): Pennsylvania underclay, near Fithian, Ill., U.S.A.; analyst O.W. Rees.
2. Whitehouse and McCarter (1958): soil Point Chevrecil, La., U.S.A.
3. Whitehouse and McCarter (1958): Recent marine mud, Atchafalaya Bay, La, U.S.A.
4. Whitehouse and McCarter (1958): Recent marine mud, Mississippi delta region, La., U.S.A.
5. Whitehouse and McCarter (1958): Recent marine mud, Gulf of Mexico.
6. Weaver (1953): weathered from feldspar in Ordovician graywacke, State College, Pa., U.S.A.; analyst G. Kunze.
7. Weaver (unpublished): Cambrian marine shale, central Texas, U.S.A.
8. Weaver (unpublished): Precambrian shale bed in Belt Limestones, Glacier National Park, Mont., U.S.A.; analyst E.G. Oslund.
9. Weaver (unpublished): Silurian, red shale in limestone, Mont., U.S.A.
10. Weaver (unpublished): Devonian, Bloomsburg Red Beds, Central Penn., U.S.A.: analyst E.G. Oslund.
11. Mackenzie et al. (1949): decomposed granite, Aberdeenshire, Scotland; analyst R.C. Mackenzie.
12. Nagelschmidt and Hicks (1943): Coal Measures shales of South Wales; analyst W.N. Adams.
13. Bates (1947): Paleozoic slate, northeastern Penn., U.S.A.; analyst R.J. Grace.
14. Bates (1947): Paleozoic slate, northeastern Penn., U.S.A. analyst R.J. Grace.
15. Mackenzie (1957b): from Upper Old Red Sandstone, Roxburgshire, Scotland; analyst J.B. Craig.
16. Grim et al. (1937): Pennsylvanian shale, near Petersburg, Ill., U.S.A.; analyst O.W. Rees.
17. Grim et al. (1937): Cretaceous shale, near Thebes, Ill., U.S.A.; analyst O.W. Rees.
18. Grim et al. (1937): slightly weathered till, Ford Col, Ill., U.S.A.; analyst O.W. Rees.
19. Grim et al. (1937): Ordovician, Mawuokita shale, Gilead, Ill., U.S.A.; analyst O.W. Rees.
20. Mankin and Dodd (1963): Silurian, Blaylock shale (alternating thin beds of shale and fine sandstone), Beavers Bend State Okla., U.S.A.,; analyst J.A. Schleicher.
21. U.S.G.S., Chinle Formation, U.S.A.; analyst W.W. Brannock.
22. Brammall et al. (1937): veneer on shear planes, Ogofau, Wales; analyst H. Bennett.
23. Maegdefrau and Hofmann (1937): in marl, probably contains interlayers of montmorillonite, Goeschwitz, Germany.
24. Maegdefrau and Hofman (1937): in kaolin deposits, Saraspatak, Hungary.
25. Gaudette (1965): in laminated pockets in Silurian dolomites, Marblehead, Wisc., U.S.A.; analyst J. Witters.
26. Gaudette et al. (1966): pockets in Ordovician limestone near Geneseo, Ill., U.S.A.; analyst J. Witters.
27. Triplehorn (1967): post-depositional in Cambro-Ordovician sandstone, Algeria.
28. Tien (1969): in hydrothermal lead–zinc ore, Silverton, Colo., U.S.A.
29. Levinson (1955): in hydrothermal kaolin deposit, St. Austell, England.

15	16	17	18	19	20	21	22	23	24	25	26	27	28	29
9.85	44.01	52.23	47.21	50.10	49.85	53.40	48.39	51.65	50.30	52.87	54.09	47.94	54.65	53.3
4.16	26.81	25.85	21.47	25.12	23.68	22.20	34.64	21.67	32.80	24.90	26.30	33.08	32.65	26.0
2.96	11.99	4.04	10.73	5.12	6.60	4.70	1.15	6.20	0.00	0.78	1.50	2.23		2.5
0.80	0.00	0.00	0.00	1.52	1.87	0.62	0.27	1.24	0.00	1.19	1.49			
3.27	2.43	2.69	3.62	3.93	1.86	2.80	0.44	4.48	1.95	3.60	2.00			
0.65	0.11	0.60	0.21	0.35	0.12	0.32	0.26	0.00	0.55	0.69	0.49	1.49	0.99	4.4
0.21	0.07	0.33	0.00	-0.05	0.34	0.25	0.22	0.31	0.52	0.22	0.22	0.34		0.2
6.31	4.78	6.56	6.17	6.93	6.64	6.90	7.82	6.08	6.72	7.98	6.87	9.48	0.09	0.3
0.62	0.64	0.37	0.00	0.50	1.40	0.70	0.11	0.00	0.00	1.02	0.68	0.32		0.01
6.22	8.08	7.88	6.17	7.18	6.80	8.20	6.07	6.44	6.98	6.73	6.89			
4.75	2.33	1.13	3.80	1.90	0.00	0.00	0.44	3.60	3.60	2.56	1.32		2.57	5.7
8.80	101.25	101.68	99.38	102.70	99.16	100.09	99.81	101.67	103.42	102.54	101.85	94.88	99.99	100.71

TABLE III

Statistical data on compositions of twenty-four illites

Variable No.	name	Mean	SE	Standard deviation	SE	Skewness	Kurtosis
1	SiO_2	49.780	0.403	1.974	0.406	−0.911	2.057*
2	Al_2O_3	26.346	0.717	3.512	0.548	0.773	0.340
3	Fe_2O_3	4.304	0.616	3.016	0.521	0.942*	0.859
4	FeO	0.611	0.142	0.694	0.068	0.686	−1.079
5	MgO	2.753	0.203	0.995	0.144	−0.485	0.012
6	CaO	0.321	0.066	0.323	0.066	1.484**	2.002*
7	Na_2O	0.245	0.043	0.213	0.071	2.623**	8.699**
8	K_2O	7.019	0.271	1.326	0.327	1.693**	3.820**
9	TiO_2	0.424	0.073	0.355	0.062	0.791	0.944
10	H_2O^+	7.120	0.241	1.181	0.157	−0.514	−0.297
11	H_2O^-	1.481	0.271	1.328	0.209	0.983*	0.370

SE − skewness = 0.472; SE − kurtosis = 0.918; *significant at 0.05 level; **significant at 0.01 level.

TABLE IV

Percent MgO in illites formed in different environments

No.	Environment	%MgO	No.	Environment	%MgO
6	weathered feldspar	1.09	2	Atchafalaya Bay mud	3.33
11	weathered granite	1.37	3	Mississippi delta, marine mud	3.32
22	veneer in shear planes	0.44	4	Gulf of Mexico, marine mud	2.90
27	post-depositional alteration of feldspar in sandstone	1.49	13	slate from marine shale	3.56
			14	slate from marine shale	3.05
			23	clay from marl	4.48

TABLE V

Illite structural formulas

	1	2	3	4	5	6	7	8	9	10	11	12	13
Octahedral													
Al	1.51	1.53	1.65	1.65	1.53	1.71	1.39	1.51	1.40	1.56	1.75	1.81	1.46
Fe^3	0.23	0.23	0.12	0.12	0.22	0.16	0.09	0.03	0.39	0.15	0.12	0.04	0.30
Fe^2	0.10	0.07	0.00	0.00	0.04	0.00	0.00	0.03	0.00	0.00	0.03	0.10	0.00
Mg	0.29	0.27	0.34	0.34	0.29	0.11	0.37	0.43	0.22	0.30	0.14	0.17	0.37
	2.13	2.10	2.11	2.11	2.08	1.98	1.85	2.00	2.01	2.01	2.04	2.12	2.13
Tetrahedral													
Al	0.55	0.54	0.66	0.66	0.50	0.53	0.38	0.57	0.62	0.54	0.60	0.78	0.68
Si	3.45	3.46	3.34	3.34	3.50	3.47	3.62	3.43	3.38	3.46	3.40	3.22	3.32
Interlayer													
Ca	0.02	0.02	0.03	0.03	0.02	0.04	0.18	0.09	0.16	0.06	0.00	0.01	0.00
Na	0.02	0.02	0.02	0.02	0.02	0.01	0.03	0.02	0.02	0.02	0.02	0.14	0.07
K	0.52	0.53	0.62	0.62	0.55	0.65	0.98	0.89	0.64	0.71	0.66	0.54	0.60
Layer charge													
Octahedral	0.00	0.04	0.01	0.01	0.09	0.17	0.82	0.46	0.19	0.27	0.05	0.09(+)	0.02(+)
Tetrahedral	0.55	0.54	0.66	0.66	0.50	0.53	0.38	0.57	0.62	0.54	0.60	0.78	0.68
Total	0.55	0.58	0.67	0.67	0.59	0.70	1.20	1.03	0.81	0.81	0.65	0.69	0.66

For legend, see Table II.

ILLITE

TABLE VI

Statistical data on structural formulas of twenty-four illites

Variable No. name		Mean	SE	Standard deviation	SE	Skewness	Kurtosis	Mean of 7 Fithian illite analyses
1	Oct. Al	1.531	0.037	0.182	0.025	0.474	−0.152	1.50
2	Fe^{3+}	0.225	0.033	0.161	0.029	1.022*	1.049	0.24
3	Fe^{2+}	0.035	0.008	0.040	0.004	0.686	−1.088	0.06
4	Mg	0.283	0.021	0.104	0.015	−0.527	−0.042	0.20
5	Tet. Al	0.600	0.023	0.111	0.019	0.568	0.939	0.44
6	Si	3.405	0.024	0.118	0.019	−0.669	0.598	3.56
7	Int. Ca	0.046	0.010	0.049	0.010	1.527**	1.967*	0.14
8	Na	0.032	0.006	0.029	0.009	2.578**	8.308**	0.05
9	K	0.615	0.024	0.118	0.029	1.751**	3.972**	0.49

SE − skewness = 0.472; SE − kurtosis = 0.918; * significant at 0.05 level; **significant at 0.01 level.

14	15	16	17	18	19	20	21	22	23	24	25	26	27	28	29
1.38	1.52	1.34	1.55	1.19	1.40	1.41	1.43	1.92	1.29	1.86	1.51	1.61	1.76	1.87	1.50
0.38	0.16	0.64	0.20	0.58	0.26	0.35	0.26	0.06	0.32	0.00	0.04	0.08	0.11		0.12
0.09	0.05	0.00	0.00	0.00	0.09	0.11	0.04	0.02	0.07	0.00	0.07	0.11	0.00		
0.32	0.34	0.26	0.27	0.39	0.40	0.19	0.30	0.04	0.46	0.19	0.36	0.20	0.15	0.10	0.43
2.17	2.07	2.24	2.02	2.16	2.15	2.06	2.03	2.04	2.14	2.05	1.98	2.00	2.02	1.97	2.05
0.64	0.49	0.89	0.49	0.62	0.60	0.53	0.46	0.74	0.46	0.69	0.46	0.43	0.82	0.55	0.51
3.36	3.51	3.11	3.51	3.38	3.40	3.47	3.54	3.21	3.54	3.31	3.54	3.57	3.18	3.45	3.49
0.00	0.10	0.02	0.09	0.03	0.05	0.02	0.01	0.04	0.00	0.08	0.10	0.07	0.05		0.03
0.02	0.03	0.01	0.04	0.00	0.01	0.05	0.04	0.03	0.04	0.07	0.03	0.03		0.02	0.04
0.53	0.57	0.43	0.56	0.53	0.60	0.59	0.68	0.66	0.53	0.57	0.70	0.58	0.80	0.72	0.69
0.10(+)0.18		0.46(+)0.21		0.09(+)0.04		0.12	0.25	0.06(+)0.11		0.04	0.49	0.31	0.09	0.19	0.28
0.64	0.49	0.89	0.49	0.62	0.60	0.53	0.64	0.74	0.46	0.69	0.46	0.43	0.82	0.55	0.51
0.54	0.67	0.43	0.70	0.53	0.64	0.65	0.89	0.68	0.57	0.73	0.95	0.74	0.91	0.74	0.79

TABLE VII

Compositional range of Fithian illite and other illites

	Fithian illite (7 analyses)	Other illites (29 samples)
Octahedral		
Al	1.38–1.61	*(1.40) 1.19–1.92
Fe^{3+}	0.21–0.37	0.00–0.64 (0.39)
Fe^{2+}	0.0–0.13	0.00–0.11
Mg	0.13–0.29	0.04–0.46 (0.43)
Tetrahedral		
Al	0.11–0.87	(0.43) 0.38–0.89 (0.82)
Si	3.89–3.13	3.11–3.62
Interlayer		
Ca/2	0.02–0.31	0.00–0.11
Na	0.02–0.07	0.00–0.14
K	0.40–0.56	(0.56) 0.43–0.98 (0.89)
Layer charge	0.55–0.88	(0.65) 0.43–1.20 (1.03)

*Values for samples with octahedral values near 2.00.

TABLE VIII

Recalculated layer charge

Sample no.	Σ	Layer charge	Interlayer cations	Reassignment Σ	Layer charge	Interlayer cations
1	2.13	0.55	0.56	2.00	0.81	0.82
4	2.11	0.67	0.67	2.00	0.89	0.89
13	2.13	0.66	0.67	2.00	0.92	0.93
15	2.07	0.67	0.70	2.00	0.81	0.81
16	2.24	0.43	0.46	2.00	0.91	0.94
23	2.14	0.57	0.57	2.00	0.85	0.85

an "end member" formula for 2M illites. This illite has a charge equivalent to that of muscovite but slightly less than half the charge originates in the octahedral sheet. The K_2O content is 10.29%. Gaudette (1965) gives a chemical analysis of a less pure (broader X-ray peaks) 2M illite from Marblehead, Wisconsin (Table II) with a K_2O value of 7.98% and a layer charge of 0.95. The Marblehead analysis and analyses of other "illites" were given by Gaudette et al. (1966); these clays showed a gradation from pure illite (no expanded layers) to an illite containing progressively more expanded layers. (The Beavers Bend, No.20, showed no detectable mixed-layering; the Marblehead, No.25, and the Rock Island, No.26, showed less than 5% mixed-layering; the Fithian, No.1, showed 10–15% mixed-layering; and "Grundite" showed 25–30% mixed-layering.) These illites, particularly the "pure" ones, may contain chloritic layers; such layers are much more difficult to identify than montmorillonitic layers.

The Ballater illite (No.11) is the only one described with a 3T structure (Levinson, 1955). Its most distinctive feature is the low content of divalent cations in the octahedral sheet, particularly Mg^{2+}. Sample No.12 is similar in composition but has a 1Md structure and also contains some kaolinite (Levinson, 1955). Samples No.6, 27, 28 and 29 are relatively pure 1M illites. The first two apparently formed from the weathering of feldspar and the latter two are of hydrothermal origin. The first three have a relatively low content of divalent cations in the octahedral sheet. On the basis of total charge and K_2O content these polytypes are similar to the 2M polytype; however, these limited data suggest that the octahedral sheet of the 2M illites (high-temperature) has more R^{2+}, largely Mg, than the other polytypic forms. Apparently R^{2+} ions are not necessary for the formation of illite and their presence may, in fact, inhibit its development. The presence of R^{2+} ions in montmorillonite may account for the relatively high temperature ($\sim 200°C$) necessary to convert it to illite. This thermal energy may be needed to allow the R^{2+} ions to migrate and form chloritic layers interspersed among muscovite type layers or it may simply be that because the large R^{2+} ions increase the size of the octahedral sheet, making it more similar in size to the tetrahedral sheet, more energy is required to force the silica tetrahedrons to rotate enough to lock in the K ions.

The relative abundance of large Fe ions in the octahedral sheet apparently determines whether a 1Md goes ultimately to a 1M or 2M polytype. Although both types of clays are dioctahedral, in the 1M iron-rich variety celadonite (Zvyagin, 1957) all three octahedral positions are of the same size (which may be true for the glauconites); whereas, in the Al-rich clays the two filled positions are smaller than the vacant position. This, in part, probably determines whether the 1M or 2M polytype is the stable phase.

The average K_2O content of these 24 illites is 7.02%. Weaver (1965) has reported that the average K_2O analyses for 16 Paleozoic and two Precambrian 2M (predominant) illites is 8.75%. Although it was necessary to correct the latter samples for non-illite impurities, it is believed that the average K_2O value is more likely characteristic of 10 Å 2M illites than that obtained from the analyses in the literature.

Frequency plots of the 001/002 ratio of 249 Paleozoic shale (Weaver,1965) and 149 soil illites (White,1962) indicate that well-ordered 10Å 2M illites have a K_2O content on the order of 9–10%. Mehra and Jackson (1959) have presented data which suggest that the completely contracted illite layers in illites have 10% K_2O. It seems likely that well-organized 10Å illite layers contain 9–10% K_2O and values less than this indicate the presence of non-illite layers or interlayer cations other than potassium.

Fig.1 and Fig.2 contain histograms showing the distribution of the oxides and the ions in their various structural positions. Most of the SiO_2, Al_2O_3, MgO, Na_2O, and K_2O have a normal type distribution. The interlayer cations have a log-normal distribution and the tetrahedral and octahedral cations have a normal-type distribution. Calculated skewness and kurtosis values are listed in Tables III and VI. The data are too limited to draw any significant conclusions. Ahrens (1954) and others have shown

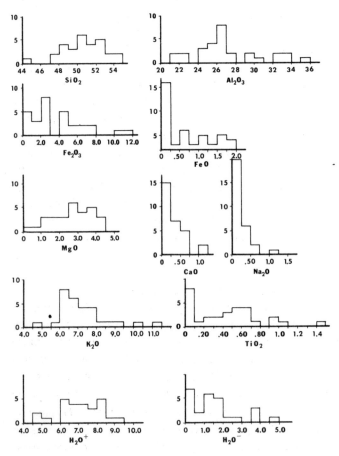

Fig.1. Histograms showing the distribution of the oxides of twenty-nine illites.

that most elements in rocks have a log-normal distribution although Shaw (1961) has indicated that normal distribution is not rare.

Tables IX and X contain correlation coefficients for the oxides and for the ions in their various structural positions. Correlations are relatively limited. There is a negative correlation between MgO and Al_2O_3 and between octahedral Mg^{2+} and octahedral Al^{3+} as would be expected. The same relation exists between ferric iron and aluminum and potassium and ferric iron. This latter correlation may reflect the fact that Fe^{3+} is the only major ion that can substitute in the illite structure and not increase the layer charge; in fact, the layer charge may decrease (assuming tetrahedral Fe^{3+} is relatively uncommon). It is also possible that some of this Fe can occur in the interlayer position. Fig.3 shows the graphical relation of K_2O and Fe_2O_3.

Several other varieties of "illite" have been described but, so far, appear to be relatively rare. Analyses of two of these are shown in Table XI. Andreatta (1949)

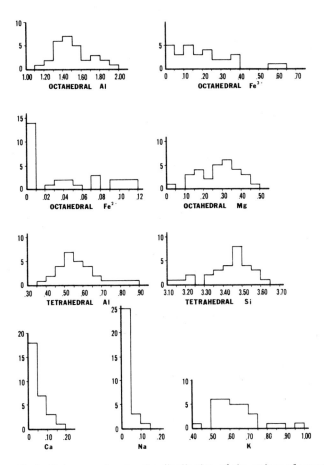

Fig.2. Histograms showing the distribution of the cations of twenty-nine illite structural formulas.

TABLE IX

Correlation coefficients for the oxides in twenty-four illites

Variable	1	2	3	4	5	6	7	8	9	10	11
No. name											
1 SiO_2	1.000										
2 Al_2O_3	−0.270	1.000									
3 Fe_2O_3	−0.461	−0.540	1.000								
4 FeO	0.145	−0.107	0.031	1.000							
5 MgO	0.155	−0.642	0.201	0.057	1.000						
6 CaO	0.244	−0.200	−0.234	−0.412	0.083	1.000					
7 Na_2O	−0.058	0.396	−0.377	0.268	−0.205	−0.113	1.000				
8 K_2O	0.351	−0.029	−0.568	−0.292	0.117	0.604	−0.094	1.000			
9 TiO_2	0.021	−0.157	0.107	0.245	−0.109	0.169	0.052	−0.058	1.000		
10 H_2O^+	−0.066	0.046	0.064	−0.015	−0.087	−0.371	0.159	−0.363	−0.166	1.000	
11 H_2O^-	−0.170	−0.269	0.206	−0.197	0.394	0.225	−0.218	−0.166	−0.280	−0.093	1.000

TABLE X

Correlation coefficients for the cations in twenty-four illites

Variable			1	2	3	4	5	6	7	8	9
No.	name										
1	Oct.	Al	1.000								
2		Fe^3	−0.776	1.000							
3		Fe^2	−0.125	0.035	1.000						
4		Mg	−0.699	0.236	0.035	1.000					
5	Tet.	Al	0.274	0.212	−0.109	−0.332	1.000				
6		Si	−0.335	−0.167	0.118	0.368	−0.945	1.000			
7	Int.	Ca	−0.057	−0.226	−0.397	0.078	−0.378	0.310	1.000		
8		Na	0.388	−0.362	0.271	−0.225	0.263	−0.229	−0.124	1.000	
9		K	0.081	−0.514	−0.300	0.133	−0.410	0.417	0.594	−0.093	1.000

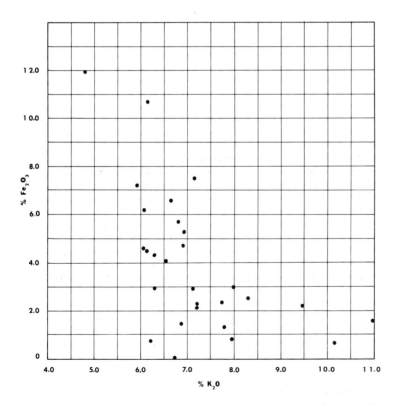

Fig. 3. Relation of Fe_2O_3 to K_2O for twenty-nine illites.

described an illite which he called illidromica of hydrothermal origin. This material is deficient in alkalis and has a high water content. It may be a mixed-layer illite-montmorillonite.

Bannister (1943) described a sodium illite, brammallite, which contains 5.22% Na_2O and 2.58% K_2O. Analysis 3 in Table XI is taken from Meyer (1935) and presumably is a sodium illite. The former sample is from a coal seam and the latter from a nodule in a conglomerate. Although sodium illite is not common, it is likely that sodium-rich layers are present interlayered with potassium-rich layers in some illites.

Trioctahedral illites have been reported by Walker (1950) and Weiss et al.(1956). Walker's analysis, which he considers only a rough approximation, is given in Table XI. The "clay biotite" occurs in a Scottish soil and is believed to be authigenic; however, it weathers so easily to vermiculite that unweathered material is difficult to find. Due to its instability, it is not likely that much clay-sized biotite exists although trioctahedral biotite-like layers may occur interlayered with dioctahedral illite layers. Such interlayering has been reported by Bassett (1959).

TABLE XI

Compositions of atypical illites

	1	2	3			1	2	3
SiO_2	53.12	40.87	47.30	Octahedral				
Al_2O_3	27.36	20.45	36.31	Al		1.67	0.89	1.84
Fe_2O_3	2.64	12.81	2.17	Fe^{3+}		0.13	0.72	0.01
FeO	1.06	0.00	–	Fe^{2+}		0.06		
MgO	2.62	6.86	–	Mg		0.26	0.72	
CaO	0.53	0.89	–	Σ		2.12	2.38	1.85
Na_2O	0.49	0.70	5.27					
K_2O	3.51	3.25	2.70	Tetrahedral				
TiO_2	0.00	2.13	–	Al		0.47	0.93	0.94
H_2O^+	8.60	11.84	5.80	Si		3.53	3.07	3.06
H_2O^-	0.00	0.00	–					
	99.93	99.80	99.55	Interlayer				
				Ca/2		0.08	0.14	
				Na		0.06	0.10	0.66
				K		0.30	0.31	0.23

1. Andreatta (1949): Illidromica, Capalbio, Italy.
2. Walker (1949): Weathered biotite, Scotland.
3. Meyer (1935): Na illite, Missouri.

The cation exchange capacity (C.E.C.) of the illite minerals is reported to range from 10 to 40 mequiv./100g; however, illites that afford values larger than 10–15 mequiv./100g usually contain some expandable layers. Ormsby and Sand (1954) showed that a good linear relation exists between C.E.C. and percent expandable layers in illites and mixed-layer illite-montmorillonites. They concluded that illite with all layers contracted would have a C.E.C. of 15 mequiv./100g.

Illite origin

Illite is stable and presumably is formed in environments where the waters have a high K^+/H^+ ratio (Garrels and Christ, 1965). Such conditions can exist either in the oceans or on the continents. However, K/Na and K/Mg ratios are equally important (low ratios favor the formation of montmorillonitic and chloritic materials) and high values are more likely to occur in continental than marine environments. At the present time more illite appears to be forming on land from the weathering of K-feldspar than is forming in the oceans. Weaver (1967) has suggested that this has been the case throughout geologic time and that there is relatively little "true" marine illite.

Yoder and Eugster (1955) and Velde (1965) have established by hydrothermal experiments that the 2M polytype is the stable form of muscovite and that the lMd polytype is metastable. Velde (1965) found that the 2M polytype is stable at temperatures as low as $125°C$ and suggested that the sequence of polytype transformation is $1Md \rightarrow 1M \rightarrow 2M$ with an increase of either time, temperature, or pressure. He concluded:

"Dioctahedral micas (alkali 'interlayer' ions having total charge near +1 per formula weight) which have either lMd or 1M polymorphs are either metastable muscovite forms or are micas with a composition differing from muscovite, e.g., glauconite, celadonite, and illite."

This suggests that 2M illites should have a muscovite composition, which does not seem to be the case in nature.

In nature the Fe-rich illites (glauconite and celadonite) appear to progress from the lMd to the 1M polytype. The Al-rich illites are predominantly the lMd and 2M varieties. If the 1M polytype is an intermediate phase, it is surprising that it is not more abundant in sediments. Recent studies of unmetamorphosed Precambrian sediments (Reynolds,1963; Maxwell and Hower,1967) have shown that the lMd polytype is relatively abundant in ancient sediments, particularly in the extremely fine fraction. The senior author has noted the relative abundance of the lMd polytype in the fine fraction of most Paleozoic rocks but has considered most of it to be mixed-layered illite-montmorillonite rather than illite. Weaver (1963a), Reynolds (1965), and Maxwell and Hower (1967) have shown that during low-grade metamorphism water is squeezed from the expanded layers and the lMd polytype is transformed into the stable 2M polytype.

Some of the 1Md material (either illite or mixed-layer illite-montmorillonite) presumably formed authigenically on the sea bottom or on land from the weathering of K-feldspars; however, much of it was formed after burial. Studies of Tertiary, Cretaceous, and Pennsylvanian thick shale sections (Weaver, 1961b) indicate that little 1Md illite was formed at the time of deposition. These shales and many others contain an abundance of expanded 2:1 dioctahedral clays with a 1Md structure, some of which is detrital and some of which formed by the alteration of volcanic material on the sea floor. With burial the percentage of contracted 10Å layers systematically increases.

More recently, Weaver and Beck (1971a) have described the diagenetic process in detail. During the first stage exchangable, inherited K migrates to the highest charged layers and inherited interlayer hydroxyl-Al and -Fe become better organized and more stable. With further burial and temperature increase the layer charge of some layers is increased by the reduction of Fe and by the incorporation of some interlayer Al into the tetrahedral sheet. K from K-feldspar and perhaps mica is adsorbed by these newly created high-charge layers increasing the amount of 10Å material. Some of the Al from the K-feldspar may move into other interlayers, increasing the number of dioctahedral chlorite layers. The resulting mixed-layer clay is composed of montmorillonite, 10Å "illite", and dioctahedral chlorite layers. These layers form below 100°C. With increasing temperature the regularity of the interlayering increases with little additional change in the proportion of layers. The interlayer material (K, Al, Fe) is apparently relatively mobile and moves to produce a layer arrangement (13Å phase) that has a high degree of stability. This phase consists of approximately equal parts of the three components.

At a temperature near 150°C kaolinite starts to decompose and the Al as hydroxy-Al moves into the interlayer position increasing the proportion of dioctahedral chlorite layers. At this stage some of the chlorite layers form packets with a sufficient number of layers to diffract as the discrete mineral chlorite. Some additional Al may move into the tetrahedral sheet at this stage and some packets of 10Å layers form (the K derived from K-feldspar). Thus, the amount of discrete 10Å "illite" and dioctahedral chlorite has increased slightly but the majority of the clay consists of a mixed-layer illite-chlorite with a lesser amount of montmorillonite.

Up to this stage of diagenesis the total chemistry is changed only slightly, if at all. Little further progress towards the muscovite-chlorite-quartz assemblage will occur unless additional ions (mainly K, Mg, Fe) are supplied to the system to balance the high Al content.

With increasing temperature (~200°C), either due to deeper burial or increase in heat-flow rates, upward migrating K, Mg, and Fe, derived from the underlying sediments, become sufficiently abundant that the remaining expanded layers are lost and some discrete 10Å illite (2M) and trioctahedral chlorite are formed; however, much of the "illite" at this stage still contains an appreciable proportion of dioctahedral chlorite and the "chlorite" contains some 10Å layers. This is the typical clay-mineral suite found in Paleozoic and Precambrian shales. Detrital mica and chlorite (trioctahedral), derived largely from metamorphic rocks, is also a sizable component of these shales.

Complete unmixing of the mixed-layered minerals requires temperatures on the order of 400°C or a relatively high degree of metamorphism (phengite → muscovite + chlorite). Velde (1964), Weaver (1965), Raman and Jackson (1966), and Weaver and Beck (1971a) have presented data to indicate chloritic layers are present in most, if not all, 10Å illites. Based on chemical data, Raman and Jackson concluded that the illites they had examined contained 20–29% chlorite layers. Weaver and Beck believe that most of the chloritic layers are dioctahedral.

The specific nature of illite is still an open question. The authors, for obvious reasons, favor the "chloritic" interpretation.

Sericite

The term sericite is frequently used to describe fine-grained dioctahedral micas. This material is usually coarser than illites and often hydrothermal in origin. Sericites

TABLE XII

Chemical analyses of some sericites

	1	2	3	4	5	6	7	8
SiO_2	45.34	46.58	46.81	46.54	49.16	49.37	52.58	46.75
Al_2O_3	31.36	37.46	36.09	30.39	30.81	29.21	23.56	32.43
Fe_2O_3	0.46	0.80	0.00	4.42	0.00	1.54	–	2.98
FeO	1.57	0.00	0.25	2.98	1.43	0.00	5.76	0.00
MgO	2.74	1.16	0.62	0.94	2.22	2.77	2.43	1.00
CaO	0.36	0.00	0.29	0.35	0.15	0.00	0.65	1.04
Na_2O	1.06	0.64	0.68	1.44	0.48	0.14	–	0.94
K_2O	9.12	6.38	10.24	6.69	10.90	9.72	9.52	5.72
TiO_2	0.19	0.00	0.01	0.00	0.04	0.00	0.18	1.15
H_2O^+	6.12	6.06	5.00	6.79	4.73	6.88	5.94	8.01
H_2O^-	1.13	0.30	0.42	0.00	0.15	0.75	–	0.00
Total	99.45	98.38	100.41	100.54	100.07	100.38	100.62	99.02

1. Minato and Takano (1952): Unnan mine, Shimane prefecture, Japan.
2. Shannon (1926): Carrol-Driscoll mine, Ida., U.S.A.
3,5.Glass (1935): pegmatite, Amelia, Va., U.S.A.
4. Doelter (1914): secondary muscovite from Epprechstein (Germany).
6. Schaller (1950): Melones, Calif., U.S.A.
7. Dana (1892): metasericite, Wildschapbackthal, Baden, Germany.
8. Carr et al. (1953): hydrous mica from Yorkshire (Great Britain) fire clay.

TABLE XIII

Sericite structural formulas

	1	2	3	4	5	6	7	8
Octahedral								
Al	1.49	1.99	1.92	1.60	1.70	1.67	1.45	1.79
Fe^{3+}	0.03	0.04		0.23		0.08	–	0.15
Fe^{2+}	0.09		0.01	0.17	0.08	–	0.33	–
Mg	0.27	0.11	0.06	0.10	0.22	0.28	0.25	0.10
	1.88	2.14	1.99	2.10	2.00	2.03	2.03	2.04
Tetrahedral								
Al	0.98	0.93	0.90	0.83	0.72	0.66	0.43	0.81
Si	3.02	3.07	3.10	3.17	3.28	3.34	3.57	3.19
Interlayer								
Ca	–	–	0.02	0.03	0.01	–	0.05	0.08
Na	0.14	0.08	0.09	0.19	0.06	0.02	–	0.12
K	0.78	0.54	0.87	0.58	0.93	0.84	0.82	0.50
Layer charge								
Octahedral	0.72	0.31(+)	0.10	0.03(+)	0.30	0.19	0.49	0.02(+)
Tetrahedal	0.98	0.93	0.90	0.83	0.72	0.66	0.43	0.81
Total	1.70	0.62	1.00	0.80	1.02	0.85	0.92	0.79

TABLE XIV

Compositional differences of some micas

	MgO	Na_2O
Illite, average	2.75	0.24
Sericite, hydro-muscovite, average	1.73	0.77
Muscovite, average*	0.63	1.15

*12 analyses from Deer et al. (1962).

may have the same chemical composition as muscovite but generally have more SiO_2, MgO, H_2O, and K_2O. As the SiO_2 content increases, sericite approaches compositions similar to the illites and is referred to as a phengite or an alurgite.

The type phengite suggested by Foster (1956) has a composition almost identical to that of the Belt illite (No.8). When the H_2O content is appreciably higher and the K_2O content lower than muscovite, the minerals have been called hydromicas or hydromuscovites. The excess H_2O in some instances is present as interlayer water, particularly in the trioctahedral hydrobiotites. Table XII contains a selection of sericite and hydromuscovite analyses and Table XIII the structural formulas. The H_2O and K_2O values of these minerals are similar to those reported for the illite minerals; however, the MgO content of the sericites and hydromuscovites is lower and the Na_2O contents higher than for the illites (Table XIV).

Radoslovich (1963b) has shown that when Na^+ ions replace K^+ ions in muscovite, the dimension of the b-axis is increased. This requires additional flattening and rotation of the silica and alumina tetrahedra. This suggests that the amount of Na^+ that can be tolerated by the mica structure increases with temperature; the increased thermal motions would allow the structure to accommodate local strains more readily. Thus, little Na^+ would be expected in low-temperature illites. In addition, Na would leach out more readily than the K and any illite that had been through the weathering stage would not retain much of its Na.

The converse is true of the Mg ion. It is more abundant in the octahedral sheets of the low-temperature 2:1 dioctahedral minerals, attaining an average value of 3.55% in the montmorillonites and even higher values in glauconite and celadonite. Mg in the octahedral position increases the size of the octahedral sheet and decreases structural strain.

Chapter 3

GLAUCONITE

Composition

Glauconite as a rock term is applied to earthy green pellets. These may be composed of a variety of clay minerals, usually iron-rich (Burst,1958; Wermund,1961). Glauconite is also used as the mineral name for the family of iron-rich 2:1 dioctahedral clays (Gruner,1935). This family contains a broad spectrum of clay minerals which are interstratified expanded and non-expanded layers in a wide range of proportions. All are disordered to some extent, those with few expanded layers showing less disorder and having a 1M structure. Disorder increases as the proportion of expandable layers increases until there is complete disorder along the "b" axis giving the 1Md polytype (Warshaw,1957; Burst,1958; Hower,1961).

A number of people have compiled data on the range and average composition of glauconite (Hendricks and Ross, 1941; Smulikowski, 1954; Borchert and Braun,1963). For the present review 69 analyses and 82 structural formulas from the literature were selected.

It is interesting to compare the average formulas obtained during two previous studies and the present one:

Hendricks and Ross (1941; 32 analyses):

$$(K, Ca/2, Na)_{0.84}$$
$$(Al_{0.47} Fe^{3+}_{0.97} Fe^{2+}_{0.19} Mg_{0.40})(Si_{3.65} Al_{0.35})O_{10}(OH)_2$$

Smulikowski (1954; 60 analyses):

$$(K_{0.67} Ca/2_{0.08} Na_{0.08})_{0.83}$$
$$(Al_{0.40} Fe^{3+}_{1.05} Fe^{2+}_{0.17} Mg_{0.41})(Si_{3.66} Al_{0.34})O_{10}(OH)_2$$

Present study (82 analyses):

$$(K_{0.66} Ca/2_{0.07} Na_{0.06})_{0.78}$$
$$(Al_{0.45} Fe^{3+}_{1.01} Fe^{2+}_{0.20} Mg_{0.39})(Si_{3.65} Al_{0.35})O_{10}(OH)_2$$

Even though the three calculations have a number of analyses in common, the similarity is more than one might expect. The slightly lower interlayer ion content in the present average formula is probably due to the inclusion of more 1Md glauconites

TABLE XV

Statistical data on composition of sixty-nine glauconites

Variable No.	name	Mean	SE	Standard deviation	SE	Skewness	Kurtosis	Range
1	SiO_2	49.215	0.317	2.633	0.458	−1.792**	6.362**	37.16−53.59
2	Al_2O_3	9.148	0.512	4.252	0.479	1.019**	1.503**	1.47−23.60
3	Fe_2O_3	17.951	0.643	5.340	0.576	−0.636*	1.213*	0.00−30.83
4	FeO	3.427	0.211	1.756	0.257	1.434**	3.889**	0.00−10.29
5	MgO	3.584	0.107	0.892	0.096	−0.095	1.186*	0.70− 5.95
6	CaO	0.644	0.106	0.884	0.217	3.531**	14.601**	0.00− 5.19
7	Na_2O	0.460	0.065	0.543	0.094	2.210**	6.277**	0.00− 3.00
8	K_2O	6.884	1.151	1.252	0.107	−0.672*	0.010	3.68− 9.01
9	H_2O^+	8.482	0.318	2.644	0.247	0.870**	0.413	4.70−15.95

SE − skewness = 0.289; SE − kurtosis = 0.570; *significant at 0.05 level; **significant at 0.01 level.

TABLE XVI

Statistical data on structural formulas of eighty-two glauconites

Variable No.	name	Mean	SE	Standard deviation	SE	Skewness	Kurtosis	Range
1	Oct. Al	0.445	0.031	0.281	0.028	1.012**	1.165*	0.00−1.27
2	Oct. Fe^{3+}	1.010	0.029	0.265	0.024	−0.192	0.649	0.35−1.77
3	Oct. Fe^{2+}	0.203	0.011	0.103	0.014	1.297**	4.371**	0.00−0.65
4	Oct. Mg	0.389	0.011	0.096	0.013	0.643*	3.718**	0.08−0.75
5	Tet. Al	0.348	0.012	0.109	0.011	0.542*	1.209*	0.11−0.72
6	Tet. Si	3.649	0.012	0.109	0.011	−0.492	1.487**	3.28−3.89
7	Int. Ca	0.065	0.009	0.077	0.015	2.525**	9.702**	0.00−0.47
8	Int. Na	0.059	0.008	0.075	0.012	2.237**	6.330**	0.00−0.42
9	Int. K	0.657	0.023	0.212	0.027	−1.940	3.210**	0.34−0.88
10	Interlayer	0.781	0.014	0.130	0.013	0.283	1.371**	0.46−1.23

SE − skewness = 0.266; SE − kurtosis = 0.526; *significant at 0.05 level; **significant at 0.01 level.

with a relatively high proportion of expanded layers. By extrapolating the data from twenty Upper Cretaceous glauconites from Bohemia, Cimbálniková (1971) concluded that a 10Å glauconite with no expanded layers should have the following composition:

$$K_{0.78}(Al_{0.45} Fe^{3+}_{1.03} Mg_{0.34})(Si_{3.65} Al_{0.31})O_{10}(OH)_2$$

This is very similar to the average formula.

The range of values along with other statistical data is shown in Tables XV and XVI. The standard deviation values and the histograms (Fig. 4,5) show the nature of the distribution of oxides and cations. Na^+ and Ca^{2+} as well as Na_2O and CaO show a

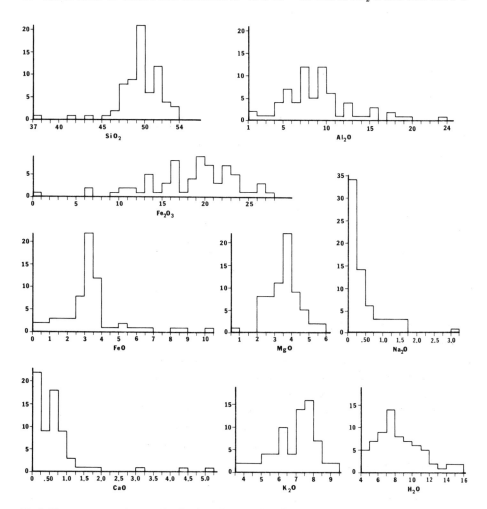

Fig.4. Histograms showing the distribution of the oxides of sixty-nine glauconite analyses.

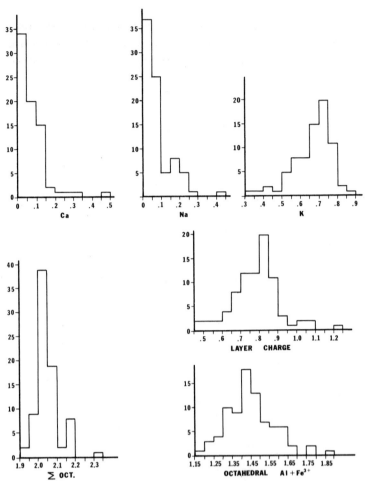

Fig.5. Histograms showing the distribution of the cations of eighty-two glauconite structural formulas.

log-normal distribution. In part, this is because a few analyses were used in which these cations were not determined. K_2O (range 3.68–9.01%) and total calculated interlayer charge (range 0.46–1.23) both show a normal distribution. The K distribution is negatively skewed with a mode at a value of 0.75–0.80. This is due to the predominance of analyses of the better ordered glauconites.

Tetrahedral Al and Si, though showing a normal-type distribution, have relatively high kurtosis values. This indicates a relatively narrow range of values for the majority of the samples (73% of the tetrahedal Al values fall in the 0.25–0.45 range). Some glauconite analyses have insufficient Al to completely fill all the tetrahedral positions and it is necessary to assign some ferric iron to the tetrahedral sheet. Whether ferric iron occurs in tetrahedral coordination or not when sufficient Al is available, has not been demonstrated.

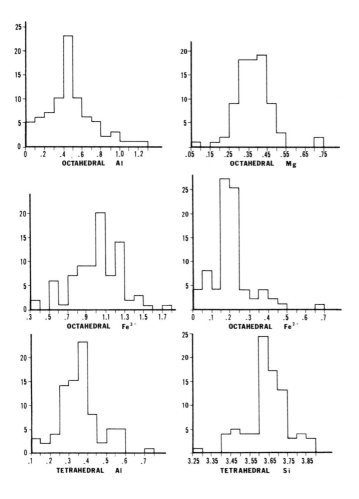

All the cations in the octahedral position have a normal-type distribution. Octahedral Al has a very distinct mode. Twenty-eight percent of the values fall in one class range (0.40–0.50) out of a total range of thirteen classes. The Fe^{2+} is even more leptokurtic having 63% of the values in the 0.15–0.25 range. Octahedral Mg also has a leptokurtic distribution but no narrow mode occurs. Eighty-nine percent of the values are in the range of 0.25–0.50.

The cation population of the octahedral sheet of glauconite includes a wider range of values than is found for most other 2:1 clays, with possible exception of the beidellites and nontronites. The average for the 82 samples is 2.054 and 77% of the values are larger than 2.00. The values range from 1.94 to 2.31. Some of these high values are due to the assignment of a portion of the interlayer cations to the octahedral sheet.

A plot (Tyler and Bailey, 1961) of the average octahedral versus interlayer cations totals for 66 analyzed glauconites and celadonites (Fig.6) shows an inverse relation. Tyler and Bailey noted that this decrease in interlayer charge tends to be compensated on the average by an increase in the total number of trivalent cations in the octahedral sheet. A possible explanation may be that some trioctahedral layers are intergrown with the dioctahedral layers. The 060 values for glauconites range from 1.51 Å to 1.53 Å as compared to a value close to 1.50 Å for Al dioctahedral 2:1 clays and close to 1.54 Å trioctahedral 2:1 clays. One explanation for this intermediate value for dioctahedral glauconite is that the Fe ion, being appreciably larger than the Al ion, stretches the layer in the b direction. As has been suggested by Warshaw (1957), Bentor and Kastner (1965), and others, the presence of some trioctahedral sheets among the dioctahedral sheets may account for this intermediate value. The presence of some trioctahedral sheets would also explain the high average octahedral population and the relatively small number of interlayer cations in those samples with a large number of trioctahedral sheets (large Σ values). Most trioctahedral minerals have some trivalent ions substituting for divalent ions in the octahedral sheet; this tends to result in a positive charge for this sheet and an overall smaller negative charge for the 2:1 layer. Consequently, fewer interlayer cations are present (biotite and saponite). There is a suggestion that those glauconites with the larger octahedral population and the larger 060 values tend to contain more expandable layers.

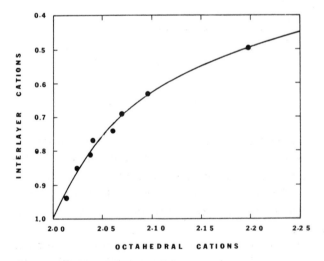

Fig.6. Variations of average interlayer cation total with average octahedral total for sixty-six analyzed glauconites and celadonites. Data divided into eight approximately equal groups. (After Tyler and Bailey, 1961.)

Table XVII contains a selection of glauconite analyses and structural formulas. No effort was made to determine the free iron oxide in most of these samples. Bentor and Kastner (1965) found 2.09% and 3.85% free iron oxide in two samples of the seven they analyzed. Apatite is a common constituent of many glauconites and, as a result, fairly high P_2O_5 values are often reported.

Glauconites on an average have approximately the same amount of interlayer K as illites and a higher calculated layer charge. The tetrahedral charge due to Al substituting for Si is considerably larger for the illites (average 0.60 for illite versus 0.35 for glauconite) and the octahedral charge is considerably less (average 0.12 for illite versus 0.47 for glauconite). For 19 of the 82 samples the tetrahedral charge exceeded the octahedral charge. Hower (1961) noted that as the octahedral charge increases there is a tendency for the potassium content to increase.

Foster (1969) divided glauconites into groups of high-potassium (greater than 0.65 K^+ per half cell) glauconites with interlayer charges greater than + 0.88 and with interlayer charges less than + 0.89 and low-potassium (less than 0.66 K + per half cell) glauconites also with interlayer charges greater than +0.88 and less than +0.89. She found that high-charge glauconites with low potassium content in interlayer positions had the remaining available interlayer positions unfilled or filled with exchangeable cations such as Na or Ca and suggested that the maximum occupancy by any exchangeable cations in the possible interlayer positions may be only 85%. She concluded that "deficiency in potassium content may be due to failure to attain maximum fixation because of change of environment, burial, uplift, or dilution, or to oxidation, with accompanying decrease in layer charge and loss of some K that had been fixed".

Divalent iron is considerably more abundant in glauconite than in illite and montmorillonite although the Mg content of glauconite is similar to that of montmorillonite. Octahedral Fe^{3+} is five times more abundant in glauconite than in illite and montmorillonite, and octahedral Al is less than one-third as abundant. The total number of trivalent ions in the octahedral position averages 1.45 as compared to 1.76 for montmorillonite and 1.68 for illite. The distribution of total trivalent ions in the octahedral sheet of glauconites is approximately normal (Fig.5). Reported values ranged from 1.15 to 1.89; however, as with the Al 2:1 clays, there is a deficiency of values less than 1.30.

In glauconites, octahedral Al ranges from 0.00 to 1.28 and Fe^{3+} from 0.35 to 1.77 (11 out of 82 samples have more octahedral Al than Fe^{3+}). Within these limits, there is a continuous isomorphous series between Al and Fe^{3+}. The composition of the basic unit cell of many glauconites is within the range of composition of the nontronite clays; however, on the average, glauconites have more Fe^{2+} and Mg and less tetrahedral Al than the average nontronite. The most variable components within the layer are the Al^{3+} and Fe^{3+} ions in the octahedral sheet. Fe^{3+} increases as Al decreases.

Foster (1969) suggested that if one started with a theoretical tetrasilicic end-

TABLE XVII

Glauconite analyses and structural formulas

	1	2	3	4	5	6	7	8	9	10	11	12
SiO_2	48.5	53.59	50.62	46.50	41.28	47.46	49.00	42.43	49.07	52.96	51.5	49.03
Al_2O_3	9.0	8.90	5.85	6.33	5.64	1.53	1.47	8.03	10.95	12.76	23.6	17.93
Fe_2O_3	20.0	15.21	15.79	14.58	25.75	30.83	24.09	28.12	15.86	13.56	10.0	13.11
FeO	3.1	3.20	2.79	2.22	3.12	3.10	10.29	0.94	1.36	2.34	3.7	1.31
MgO	3.7	3.37	4.85	5.95	5.41	2.41	2.91	2.85	4.49	4.11	2.00	2.79
CaO	0.4	0.06	0.81	4.45	0.67	—	0.19	0.14	0.07	—	0.06	0.39
Na_2O	1.5	0.21	0.13	0.42	0.25	—	trace	0.66	0.13	0.47	0.62	0.10
K_2O	6.1	7.09	7.19	6.00	3.79	7.76	5.29	6.27	7.51	8.69	4.8	7.84
TiO_2	—	0.15	trace	0.10	0.19	—	0.80	0.25	0.15	—	—	1.06
H_2O^+	7.3	7.52	19.12	—	14.46	7.00	7.36	5.93	6.63	4.91	4.7	6.00
	—	1.73		—				4.15	3.66			
Total	99.6	101.03	107.14[a]	86.55[b]	100.56	100.09	101.40	99.77[c]	100.20[d]	99.80	100.98	99.93[e]
Octahedral												
Al	0.35	0.63	0.41	0.37	0.00	0.00	0.00	0.05	0.60	0.74	1.26	0.99
Fe^{3+}	1.12	0.82	0.91	0.89	1.34	1.77	1.21	1.66	0.89	0.71	0.53	0.70
Fe^{2+}	0.19	0.19	0.18	0.15	0.28	0.20	0.65	0.07	0.09	0.14	0.23	0.08
Mg	0.40	0.36	0.55	0.72	0.75	0.28	0.33	0.33	0.49	0.43	0.20	0.30
Σ	2.06	2.00	2.05	2.13	2.37	2.03	2.19	2.11	2.07	2.01	2.15	2.07

COMPOSITION

	1	2	3	4	5	6	7	8	9	10	11	12
Tetrahedral												
Al	0.42	0.13	0.12	0.23	0.56	0.36	0.13	0.68	0.36	0.31	0.58	0.51
Fe^{3+}	—	—	—	—	0.16	—	0.16	—	—	—	—	—
Si	3.58	3.87	3.88	3.77	3.28	3.64	3.71	3.32	3.64	3.64	3.42	3.49
Interlayer												
Ca/2	0.06	0.00	0.00	0.00	0.00	—	0.02	0.02	0.04	—	0.01	0.03
Na	0.21	0.03	0.02	0.07	0.15	—	—	0.10	0.04	0.06	0.08	0.01
K	0.57	0.65	0.70	0.62	0.50	0.76	0.51	0.62	0.72	0.77	0.40	0.71
Calculated layer charge	0.83	0.68	0.72	0.69	0.65	0.76	0.71	0.75	0.73	0.83	0.49	0.75
% Expanded layers	10	10	20	20	50	—	10	10	—	—	20	10

[a]$P_2O_5 = 1.42$, $CO_2 = 0.86$, $S = 0.08$; [b]$P_2O_5 = 3.90$, $CO_2 = 0.81$, $S = 0.07$; [c]$P_2O_5 = 0.15$, $MnO = 0.08$; [d]$P_2O_5 = 0.19$, $Cr_2O_3 = 0.07$; [e]$P_2O_5 = 0.37$

1. Cambrian, Franconia Formation (sandstone), Norwalk, Wisc., U.S.A.: Schneider (1927); analyst T.B. Brighton.
2. Upper Cretaceous, Venezuela (silty chert): Warshaw (1957); analyst J.A. Solomon.
3,4. Eocene, N.J., U.S.A. (sandy opaline chert): Warshaw (1957); analyst M.J. Crooks.
5. Quaternary, Gulf of Mexico (fillings of Foraminifera): Warshaw (1957); analyst J.A. Solomon.
6. Pacific Ocean, off California, U.S.A.: Collet and Lee (1906).
7. Cretaceous secondary glauconite after Precambrian iron-formation, Auburn mine, Minn., U.S.A.: Tyler and Bailey (1961).
8. Lower Cretaceous, Makhtesh Hatsera, Israel (calcareous argillaceous sandstone): Bentor and Kastner (1965); analyst M. Gaon.
9. Upper Cretaceous or Tertiary, Milburn Table Hill, Otago, New Zealand (limestone): Hutton and Seelye (1941); analyst F.T. Seelye.
10. Lower Siberian of Udriass (limestone): Glinka (1896).
11. Subsurface, California, U.S.A., Eocene (sandstone): Burst (1958).
12. Upper Jurassic, Morrison Formation, Blue Mesa, Uravan, Colo., U.S.A. (continental sandstone): Keller (1958); analyst B. Williams.

member and substituted Al for Si in the tetrahedral sheet and R^{3+} for R^{2+} in the octahedral, the sum of the decrease in Si^{4+} and R^{2+} should equal the sum of the increase in Al^{3+} and R^{3+}. In those samples where the decrease in R^{2+} was greater than the decrease in Si, and the increase in R^{3+} was greater than the increase of Al in the tetrahedral sheet, she postulated oxidation of ferrous to ferric iron in the octahedral sheet. This oxidation would have the effect of decreasing the layer charge, thus resulting in low-potassium, low-charge glauconites.

Frequency-distribution curves of the ions in glauconite and illite show overlap for each ion. The minimum overlap is in octahedral Al. The maximum amount of Al in the octahedral sheet of glauconite is 1.27 and the minimum amount in illite 1.19. The two populations are mutually exclusive and, as such, are more characteristic of the two families of minerals than is the iron content. Ferrous iron is almost as selective as Al. On the basis of the average data for the major 2:1 clay groups, it appears that if the small Al ion is the dominant ion in the octahedral sheet (1.53 for illites and 1.49 for montmorillonites), it comprises a larger portion of the octahedral cations than when the larger ferric ion is dominant (1.01 for glauconite). The remaining octahedral positions in illite and montmorillonite are filled with the large ferric and Mg ions with Mg being slightly more abundant. In glauconites, on an average, half of the remaining positions are filled with the relatively small Al ion and half with the larger Mg ion. The frequency distributions of these two ions are similar and it is apparent that it is necessary for both to be present in order to minimize strain.

There has been considerable effort by a number of people to show correlations among the various chemical components of glauconite and among composition and crystal structure, age, environment, lithology, color, morphology, etc. Manghnani and Hower (1964) showed that there is a good negative relation between the amount of potassium and the percentage of expandable layers (Fig.7). Glauconites are similar to the Al-rich mixed layer clays because the larger the proportion of contracted 10Å layers, the greater the density of potassium in the interlayer position. However, the potassium density between the contracted glauconite layers is consistently less than that of the Al-rich mixed-layer clays (see Fig.18) even though the total layer charge is usually larger. In part, this may be due to the tendency of glauconites to have a larger proportion of the charge originating in the octahedral sheet. When the charge originates in the tetrahedral sheet, not only is it closer to the potassium ion, but the charges are localized and will tend to attract K^+ to the immediate vicinity of the charges. The charges originating in the octahedral sheets would be less specific and more diffuse in their effect and could be statistically satisfied by a smaller number of K^+ than the point-like charges originating in the tetrahedral sheets. Köster (1965) plotted percent MgO versus percent K_2O to show a direct positive relation between the amount of Mg in the octahedral sheet and the amount of interlayer K. Cimbálinková (1971) showed that as the proportion of expanded layers increased the tetrahedral charge decreased. This demonstrates the montmorillonitic nature of the

COMPOSITION

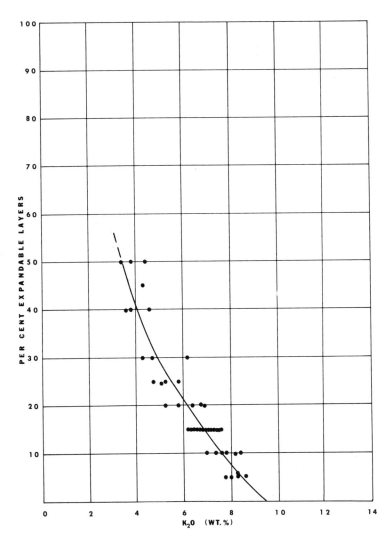

Fig.7. The relation between potassium content and percent expandable layers in glauconite. (After Manghnani and Hower, 1964.)

expanded layers. It is of interest to note that the minimum K_2O content of the contracted layers is 7%, the same minimum that exists in the Al 2:1 contracted layers. The maximum K_2O value for glauconites is 8.7% and for Al 2:1 contracted layers 9.9% K_2O.

Most of the other relations that have been suggested are not, as yet, well-founded. Tables XVIII and XIX show the correlation coefficients for the oxides and the ions in their various structural positions. There are few correlations execpt those that are forced: for example, octahedral Fe^{3+} negatively correlated with octahedral Al and

TABLE XVIII

Correlation coefficients for the oxides of sixty-nine glauconites

No.	Variable name	1	2	3	4	5	6	7	8	9	10
1	SiO_2	1.000									
2	Al_2O_3	0.223	1.000								
3	Fe_2O_3	−0.401	−0.799	1.000							
4	FeO	0.155	−0.058	−0.240	1.000						
5	MgO	−0.040	−0.128	−0.092	−0.088	1.000					
6	CaO	−0.526	−0.095	−0.013	−0.005	0.356	1.000				
7	Na_2O	−0.066	0.101	−0.024	−0.116	0.038	0.215	1.000			
8	K_2O	0.328	−0.045	−0.095	0.138	−0.181	−0.264	−0.441	1.000		
9	TiO_2	−0.009	−0.061	0.097	0.224	0.040	−0.012	−0.142	−0.221	1.000	
10	H_2O^+	−0.568	−0.048	−0.003	−0.263	0.141	0.270	−0.021	−0.496	−0.058	1.00

TABLE XIX

Correlation coefficients for the cations in eighty-two glauconites

No.	Variable name	1	2	3	4	5	6	7	8	9	10
1	Oct. Al	1.000									
2	Oct. Fe^{3+}	−0.895	1.000								
3	Oct. Fe^{2+}	−0.241	−0.053	1.000							
4	Oct. Mg	−0.130	−0.098	−0.068	1.000						
5	Tet. Al	0.066	0.142	−0.158	−0.124	1.000					
6	Tet. Si	−0.044	−0.162	0.159	0.126	−0.979	1.000				
7	Int. Ca	0.070	−0.192	0.133	−0.080	0.172	−0.176	1.000			
8	Int. Na	−0.008	0.008	−0.128	0.068	−0.256	−0.264	0.340	1.000		
9	Int. K	0.053	−0.118	−0.013	0.015	−0.026	0.020	0.186	−0.025	1.000	
10	Layer charge	−0.126	−0.009	0.043	−0.069	0.026	−0.031	0.699	0.378	0.359	1.00

tetrahedral Si negatively correlated with Al. Ca is positively correlated with layer charge and CaO is negatively correlated with SiO_2. If the Ca is present as an exchangeable cation, as it is usually assumed to be, the relation to the interlayer charge is difficult to explain. Much of the Ca may be present in apatite; however, the almost universal presence of Ca in the 2:1 clay minerals suggests some of it may be as firmly fixed in the interlayer position as K. H_2O is negatively correlated with SiO_2 and K_2O. This presumably means that glauconites with a lower tetrahedral charge and with fewer contracted layers have more structural hydroxyls. It is more likely that H_2O^- (interlayer water), which is more abundant in the low-K glauconites, is not completely accounted for in many of these analyses and is, in part, included with the H_2O^+ water.

Hower (1961) suggested that there is a relation between Fe_2O_3 and K_2O, which he believes implies that both were adsorbed from sea water on to a degraded layer. There is nothing in the present data to suggest that this relation exists. The studies of Ehlmann et al. (1963) and Seed (1965) indicate that in the initial stages of formation, glauconite pellets are nearly white and composed largely of expanded layers. The potassium content of this material is low (4%) but the iron content is 20% or more. Associated with these white pellets are light green and dark green pellets. The potassium content and the proportion of contracted layers both increase in the darker green pellets although the iron content remains nearly constant. It is believed that these three types of iron-rich 2:1 clays represent stages in the diagenesis of glauconite on the ocean floor. However, Ehlmann et al. noted that the morphology of all three types was different and there was no direct evidence to indicate the three types represent an evolutionary sequence. In any event, it appears that the iron is present in the early stages of lattice development, regardless of how it got there, and that potassium is attracted later.

Since this review was originally completed, Foster (1969) published a review in which similar conclusions are drawn about the glauconites and celadonites. The lack of correlation between iron and potassium content in glauconite is substantiated in her paper. Foster considered the process of glauconitization to be "of two separate, unrelated processes, incorporation of iron into the crystal structure and fixation of potassium in interlayer positions, with incorporation of iron and development of negative layer charge preceding complete fixation of potassium".

The various green colors of glauconite are apparently due to variations in the Fe^{3+}/Fe^{2+} ratio (Shively and Weyl, 1951; Warshaw, 1957). The greenish-white and light olive glauconites have ratios on the order of 9–10 and those of the bluish-green variety have ratios in the range of 2–5. The information in the literature suggests that there is a tendency for the darker green glauconites to have a higher proportion of contracted 10 Å layers. One might suspect that the darker green glauconites, that have a relatively high Fe^{2+} content, would have a higher K_2O content. A plot of Fe^{3+}/Fe^{2+} versus K_2O suggests only a very vague negative relation.

Bailey and Atherton (1969) studied glauconite pellets varying in color from grass green through a mottled pale and grass green to pale green. The grass green pellets were the mineral glauconite; the pale green pellets were mixtures of francolite (fluorapatite) intergrown with the glauconite; and the mottled pellets reflected a less homogeneous dispersion of glauconite in the mixture.

Cation exchange capacities as determined by Manghnani and Hower (1964) show a good linear relation to the relative proportion of expanded layers (Fig.8). The C.E.C. ranges from 5 to 12 mequiv./100 g for those glauconites having only 5% expandable layers and increases to 30–40 mequiv./100 g for those having 40–50% expandable layers. The C.E.C. of the expanded layers alone ranges from 60 mequiv./100 g for those clays with a relatively high proportion of expandable layers to 90 mequiv./100 g for those with only a few expandable layers. These values are lower than the values for typical dioctahedral montmorillonites. It is quite likely that the relatively low C.E.C. of the expandable layers of glauconite is due to the presence of some trioctahedral layers and hydroxy interlayer material.

Ca appears to be slightly more abundant than Na in the exchange positions (Tables XV and XVI) although Foster (1969) found Na to be more common for some glauconites. Analyses reported by Owens and Minard (1960) show that Mg can also be a major exchange cation (Table XX).

Electron micrographs show that glauconites have both a subequant and a lath-shaped morphology. Limited data suggest that the lath-shaped glauconites have few expanded layers, high layer charge, and a relatively low octahedral population.

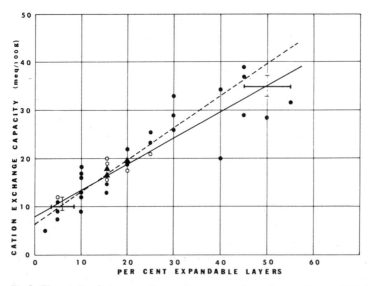

Fig.8. The relation between cation exchange capacity and percent expandable layers in glauconites. Solid dots represent one sample; open circles two samples; dashed line represents trend excluding points with > 30% expanded layers; solid line represents C.E.C. (mequiv./100 g) = 8.2 + 0.54 (% expanded layers). (After Manghnani and Hower, 1964.)

TABLE XX

Exchange capacity and exchangeable cations of glauconite concentrates from N.J. Sewell (After Owens and Minard, 1960)

Formation	Sieve size	Mequiv./100 g					Total K_2O in sample (%)
		Na	K	Ca	Mg	total (sum)	
Hornerstown	+35	0.06	0.03	15.4	9.8	25.3	7.4
Hornerstown	+60	0.12	0.02	15.4	7.0	22.4	–
Navesink	+35	0.05	0.02	15.8	4.9	20.8	7.2
Navesink	+60	0.03	0.02	15.4	5.6	21.0	–
Navesink	+120	0.07	0.02	18.5	7.7	26.3	–

Age

Smulikowski (1954) and Hower (1961) have suggested that the older glauconites have more K_2O than the younger glauconites. Warshaw (1957) and Bentor and Kastner (1965) found no change with age. Hurley et al. (1960) and Evernden et al. (1961) made K-Ar determinations on a large number of glauconites and both suggested that the K content increased with age. Their K and age data are plotted in Fig.9, along with data from three Recent samples. If a trend exists, it is extremely vague. Such a tenuous relation suggests that, if there is indeed any relation at all, the K content varies primarily with some factor other than age and the relation to age is only second-order. Sampling is far from random. For example, all of the K values (6) greater than 6.6% K reported by Hurley et al. (1960) are from the Franconia Formation.

Hurley et al. showed that the age measurements on glauconite by the K-Ar method indicate a consistent variation with geologic age although the results are 10–20 % short of age measurements on micas. With suitable techniques and a careful selection of samples Evernden et al. (1961) believe it is possible to obtain ages as accurate as those from igneous biotite. Both studies show that deeply buried glauconites lose ^{40}Ar and the resulting age is too young. They further suggest that K is continuously added or fixed in glauconite (contracting some of the expanded layers) throughout geologic time. At the same time ^{40}Ar is lost from the contracting layers as the water is removed from between the layers. The basis for this reasoning are the data in Fig.9 which at best indicate that on an average the Cambrian glauconites, analyzed to date, have slightly more K than the Tertiary glauconites, but not appreciably more than the Cretaceous glauconites. Hurley et al. (1960) also state that "older

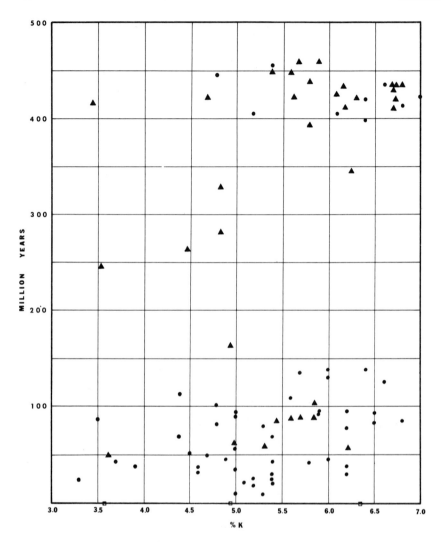

Fig.9. The relation between K content of glauconite and geologic age. (Dots from Evernden et al., 1961; triangles from Hurley et al., 1960; squares from Ehlmann et al., 1963.)

glauconites contain 10–20% less expandable layers than the younger ones" (10% versus 20–25%). Again, this is an average value and the data of Warshaw (1957), Hower (1961), Ehlmann et al. (1963), Bentor and Kastner (1965), and Seed (1965) indicate that glauconites having less than 10% expandable layers are relatively common throughout the geologic column from Cambrian to Recent. This is also evident from the comparison of the excellent relation between potassium content and percent expandable layers (Fig.7) and the relation between potassium and age shown in Fig.9.

If glauconites afford correct ages or ages which are consistently 10–20% low, then it is unlikely that there has been any systematic juggling of the K and Ar content of glauconites throughout geologic time. Further, studies of recent glauconites (Ehlmann et al., 1963; Porrenga, 1966) suggest that glauconites obtain most of their potassium very rapidly (few years or hundreds of years) after formation of the pellet-shaped mass. During the initial stages of glauconite formation, the white clay pellets contained 2–4% K_2O. Associated dark green pellets which were older, but presumably formed by modification of the white pellets, contained 7–8% K_2O.

Conway (1942) and Smulikowski (1954) have interpreted the potassium-distribution data to indicate that there is a potassium deficiency in Tertiary and Recent seas. However, Spiro and Gramberg (1964) made analyses of the composition of cations adsorbed on argillaceous rocks of northern Siberia and concluded that "...the highest content of potassium is inherent in marine water of the Permian Period. In Triassic seas the content of the potassium dropped significantly, and reached a minimum in seas of the Jurassic Period. Beginning with Cretaceous, the amount of potassium in sea water increased again, and during the Quaternary Period its level approached that of the Permian seas." These ideas are extremely speculative.

Table XXI, after Smulikowski (1954), shows the average composition of glauconites divided on the basis of age and lithology. Smulikowski concluded:

"The older the geological formation, the smaller in its glauconite the prevalence of ferric iron over aluminum in the octahedral layer and the greater in its glauconite the total amount of interlayer cations. On the other side, the greater the lime carbonate content in the sediment, the lower the sum of octahedral cations in the glauconite, the smaller the prevalence in it of ferric iron over aluminum in the octahedral layer, and the higher the proportion of interlayer cations."

As Smulikowski points out, there is a direct relation between the geologic age and lime content. Six of ten Early Paleozoic, six of thirteen Creaceous, four of eighteen Tertiary and no Recent glauconites are associated with carbonate rocks. He suggests: "... the average analyses of glauconites set in the sequence of growing geological age, may show the increasing role of limestone".

Warshaw (1957) and Hower (1961) noted a relation between lithology and character of glauconite. The better crystallized, more pure glauconites with few expandable layers tend to occur in clean sandstones, dolomites and limestones, whereas those with a large proportion of expandable layers and a mixed mineralogy tend to occur in argillaceous sandstone and marls. This relation may in part be due to differing rates of sedimentation ("... mixed-layering is believed to result from incomplete degradation of pre-existing silicates" – Warshaw, 1957); however, competition for available potassium and the presence of other clay minerals may be important. The clay content of a sediment is in part controlled by the environment of deposition but the source area is also important. The lack of clay in a sediment may imply that there are few clays in the source area or that weathering has proceeded through the clay stage to the oxide and gel stage. Thus, there is an excellent chance that source material for glauconite in a quartzite–limestone sequence would differ from that in a relatively clay-

TABLE XXI

Average compositions and structural formulas of selected glauconite suites (after Smulikowski, 1954)

	h	t	k	s	p	g	m	w	T
SiO_2	50.19	49.40	50.21	49.75	49.91	48.92	48.82	50.09	49.58
Al_2O_3	4.83	8.04	8.17	10.04	7.27	7.15	9.13	9.87	8.41
Fe_2O_3	24.83	19.20	17.95	17.44	20.39	20.50	18.37	17.06	18.81
FeO	2.47	2.62	3.18	2.55	2.45	3.20	2.29	2.31	2.78
MgO	3.06	3.39	3.51	3.55	3.54	3.26	3.66	3.40	3.75
K_2O	6.50	7.25	7.26	7.75	7.21	7.38	6.41	7.66	7.11
Na_2O	0.51	0.39	0.59	0.66	0.47	0.44	1.13	0.64	0.55
CaO	0.49	0.35	0.47	0.57	0.24	0.38	0.79	0.48	0.51
H_2O	7.27	9.43	8.01	7.39	8.28	8.41	8.55	8.43	8.12
Total	100.15	99.77	99.35	99.70	99.76	99.64	99.15	99.94	99.62
Octahedral									
Al^6	0.133	0.399	0.426	0.501	0.333	0.300	0.433	0.530	0.398
Fe^{III}	1.381	1.077	1.000	0.957	1.140	1.158	1.027	0.943	1.048
Fe^{II}	0.153	0.164	0.196	0.156	0.151	0.200	0.143	0.141	0.171
Mg	0.338	0.377	0.387	0.386	0.392	0.360	0.406	0.370	0.413
Tetrahedral									
Al^4	0.289	0.310	0.285	0.364	0.299	0.331	0.371	0.325	0.335
Si	3.711	3.690	3.715	3.636	3.701	3.669	3.629	3.675	3.665
Interlayer									
K	0.613	0.691	0.684	0.720	0.681	0.707	0.607	0.714	0.670
Na	0.071	0.058	0.084	0.097	0.071	0.063	0.161	0.088	0.080
Ca	0.040	0.027	0.036	0.044	0.018	0.032	0.062	0.040	0.040

h = Recent glauconite (average from 6 analyses).
t = Tertiary glauconite (average from 18 analyses).
k = Cretaceous glauconite (average from 13 analyses).
s = Early Paleozoic glauconite (average from 10 analyses).
p = glauconite from sands (average from 17 analyses).
g = glauconite from sandstones (average from 9 analyses).
m = glauconite from marls (average from 6 analyses).
w = glauconite from limestones (average from 11 analyses).
T = typical glauconite (average from 22 analyses).

TABLE XXII

K content (in %) in glauconites from Franconia Formation

5.08	5.17	6.29	6.53	6.63	6.67	6.68	6.68	6.68
6.69	6.72	6.73	6.79	6.83				

rich depositional sequence. It is believed that most of the authigenic marine Al illites are associated with the quartzite–limestone facies (Weaver, 1964a). Hower (1961) suggested that the potassium content of glauconites may be primarily a function of lithology which, in turn, is partially related to age. This explanation is more realistic than that of continuous diagenesis.

Too few chemical analyses are available of glauconites from one formation to provide much indication of the amount of variability at the formation level. However, Table XXII lists 14 potassium analyses of glauconites from the Upper Cambrian Franconia Formation of Minnesota and Wisconsin.

The values have a well-defined mode but the extremes cover a range of over 2% K_2O. This is a wide range, particularly in view of the mineral homegeneity and the well-crystallized nature of the Franconia glauconites. The range of values for these well-crystallized glauconites of a restricted age (411–450 m.y.; Hurley et al., 1960) is as great as the variability from the Recent to the Cambrian.

Origin

The conditions favoring the formation of glauconite have been discussed by most of the authors referenced in this review and will only be summarized here.

The most favorable physical environment is one of slow deposition, agitated water and paucity of clay minerals — though such an environment is far from critical. Chemically, a saline, oxidizing environment is favored. Sufficient organic material is necessary to create local reducing conditions. Higher than normal amounts of iron and potassium are required although potassium is probably less restrictive than iron. Sufficient potassium is present in the sea and in the organic material producing the reducing conditions. Magnesium is also an essential constituent but is likewise readily available in a normal marine environment.

Most glauconite appears to have originated as faecal pellets or clay filling of fossil tests; however, it occurs replacing biotite, heavy minerals, feldspar, muscovite, quartz, rhyolite, and volcanic glass. The parent material (in the faecal pellets and tests fillings) is thought to be a silica-alumina gel (Takahashi, 1939), ferri-alumino-siliceous hydrogel (Smulikowski, 1954), or a degraded 2:1 layer (Burst, 1958). It is likely that all of these and more can be converted to glauconite under suitable conditions. Iron may be present in the original starting material or adsorbed by it rapidly, at least more rapidly than the potassium (Ehlmann et al., 1963; Seed, 1965; Porrenga, 1966). Most of this iron is initially in an oxidized state and may be inherited from degraded biotite or nontronite. Colloidal iron, iron-adsorbed on the surface and interlayer position of clay minerals and on quartz, and organo-iron complexes are presumably other major sources. Mildly reducing conditions created by decaying organic matter and bacterial action apparently favor the formation of a magnesium-rich nontronite layer with a relatively low layer charge. Additional reduction of the iron occurs when the layer charge is increased and more potassium is adsorbed causing layers to contract to 10 Å. If all the organic matter is destroyed and the environment becomes oxidizing, it is possible for some of the ferrous iron to revert to the ferric state and for some potassium to be released. Conversely, if Eh increases after burial, the layer charge may continue to increase and additional potassium will be attracted from sites where ΔF is less than ΔF in the glauconite structure.

The relatively narrow compositional range and the near normal distribution of most of the ions (Fig.5) suggest that, regardless of the starting material, the allowed variation in the chemical environment is quite restricted.

Non-marine glauconites

Dyadchenko and Khatuntzeva (1955) and Keller (1958) give analyses of non-marine glauconites altered from montmorillonite and feldspar, respectively.

Dyadchenko and Khatuntzeva:

$$K_{0.69}Na_{0.10}Ca_{0.06}$$
$$(Al_{0.13}Fe^{3+}_{1.21}Fe^{2+}_{0.20}Mg_{0.47})(Si_{3.52}Al_{0.42}Ti_{0.06})O_{10}(OH)_2$$

Keller:

$$K_{0.71}Na_{0.01}Ca_{0.03}$$
$$(Al_{0.99}Fe^{3+}_{0.70}Fe^{2+}_{0.08}Mg_{0.30})(Si_{3.49}Al_{0.51})O_{10}(OH)_2$$

TABLE XXIII

Non-marine glauconite-illite

	Parry and Reeves (1966)*	Porrenga (1968)**
SiO_2	41.6	54.18
Fe_2O_3	16.1	12.11
FeO	0.7	2.03
Al_2O_3	11.0	14.20
MgO	4.8	4.01
CaO	0.9	0.68
K_2O	3.8	6.07
Na_2O	3.3	0.04
TiO_2		0.22
MnO		0.03

$$\phantom{*(Ca_{0.05}Na_{0.51}K_{0.39})}\overset{+1.06}{}\quad \overset{-0.45}{}\quad \overset{-0.61}{}$$
*$(Ca_{0.05}Na_{0.51}K_{0.39})(Al_{0.44}Fe^{3+}_{0.99}Fe^{2+}_{0.05}Mg_{0.58})(Si_{3.39}Al_{0.61})O_{10}(OH)_2$
$$\phantom{**(Ca_{0.05}Na_{0.01}K_{0.53})}\overset{+0.64}{}\quad \overset{-0.38}{}\quad \overset{-0.26}{}$$
**$(Ca_{0.05}Na_{0.01}K_{0.53})(Al_{0.89}Fe^{3+}_{0.63}Fe^{2+}_{0.12}Mg_{0.41})(Si_{3.74}Al_{0.26})O_{10}(OH)_2$

Parry and Reeves (1966) reported a lustrine glauconitic mica that has a structural formula intermediate between glauconite and illite. Porrenga (1968) also reported a clay mineral of an intermediate nature (Table XXIII).

These non-marine "glauconites" represent a wide range of compositions, but the samples of Keller, Porrenga, and Parry and Reeves have compositional features that would place them between glauconite and illite. Porrenga (1968) plotted the number of Al^{3+} ions in tetrahedral position against the number of Fe^{3+} ions for these three glauconite micas (or illites) in addition to illites and glauconites to show the intermediate nature of these samples. (The sample of Dyadchenko and Khatuntzeva would plot in the glauconite range of the graph.) Porrenga pointed out that these three clay minerals are believed to have formed by the alteration of detrital illite and/or montmorillonite and that these clays are "lagoonal, lacustrine or fluviatile in origin as opposed to the pelletoidal glauconite which is marine in origin".

Chapter 4

CELADONITE

The origin and composition of celadonite has been discussed by Hendricks and Ross (1941) and more recently by Wise and Eugster (1964) and Foster (1969). Celadonite is non-marine in origin, commonly occurring as fillings in vesicular cavities or directly replacing olivine basalts. "Volcanic clastic rocks upon diagenetic alteration or low-grade metamorphism (zeolite facies) often develop celadonite-bearing assemblages" (Wise and Eugster, 1964). Shashkina (1961) described the formation of celadonite veins in slightly weathered basalt by the process of "joint coagulation of Fe_2O_3, Al_2O_3, and SiO_2 from colloidal solutions saturated with alkali". Celadonite occurs only as the 1M polytype.

The 15 analyses in Table XXIV are selected from those reported by Wise and Eugster (1964). They summarize:

"Analysis No. 1 represents the only truly tetrasilic celadonite. The average value for Si in the 17 analyses is 3.83. Except for Nos. 17–19 variations are between 3.73 and 4.00. Octahedral occupancy is close to 2 for all celadonites, with an average of 2.05. Interlayer cations often number substantially less than 1.0 and average 0.81 with a low value of 0.38 (No. 14). This, of course, is typical also for glauconites and hydrous micas."

Analyses reported by Radonova and Karadzhova (1966) approach the composition of the tetrasilicic end-member having only 0.01–0.03 Al in the tetrahedral sheet; however, the octahedral charge is only approximately 0.8 rather than 1.0. These clays were formed by low-temperature hydrothermal alteration of porphyritic rocks.

Shashkina (1961) gives analyses of four samples from the Volynya Basalt which lie within the range of values in Table XXIV:

$$K_{0.30-0.75}Na_{0.02-0.08}Ca_{0.03-0.08}$$
$$(Al_{0.32-0.78}Fe^{3+}_{0.42-0.69}Fe^{2+}_{0.14-0.23}Mg_{0.93-0.92})(Al_{0.10-0.24}Si_{3.76-3.90})$$
$$O_{10}(OH)_2$$

Celadonites commonly have 0.00–0.30 tetrahedral Al (Fig. 10) whereas most glauconites have between 0.25 and 0.60 (Fig. 5). The smallest value reported for glauconite is 0.11 but 23 of the 82 glauconite analyses collected have less than 0.30 tetrahedral Al. Thus, there is considerable overlap and the composition of this sheet can not be considered exclusively diagnostic. It is more likely that there is a continuous series with two well developed modes.

TABLE XXIV

Chemical analyses and structural formulas of some celadonites (After Wise and Eugster, 1964)

	1	2	3	4	5	6	7
SiO_2	55.61	53.23	54.30	54.73	52.69	55.30	50.70
TiO_2	–	–	–	–	–	–	–
Al_2O_3	0.79	2.13	5.08	7.56	5.79	10.90	4.72
Fe_2O_3	17.19	20.46	14.77	13.44	9.75	6.95	15.34
FeO	4.02	4.14	4.82	5.30	5.37	3.54	2.00
MgO	7.26	5.67	6.05	5.76	8.54	6.56	9.32
MnO	0.09	–	0.09	–	0.31	–	trace
CaO	0.21	–	0.80	0.00	1.16	0.47	1.32
Na_2O	0.19	–	3.82	–	0.39	0.00	0.29
K_2O	10.03	7.95	4.85	7.40	6.21	9.38	4.44
Li_2O	–	–	–	–	–	–	–
H_2O^+	4.88	6.18	5.64	6.40	10.48	6.51	12.52
H_2O^-							
CO_2	–	–	–	–	–	–	–
P_2O_5	–	–	–	–	–	–	–
Total	100.27	99.76	100.22	100.59	100.69	99.61	100.65
Tetrahedral							
Si	4.00	3.90	3.86	3.87	3.88	3.88	3.79
Al	–	0.10	0.14	0.13	0.12	0.12	0.21
Octahedral							
Al	0.07	0.08	0.28	0.50	0.38	0.78	0.21
Fe^{3+}	0.93	1.13	0.79	0.71	0.54	0.37	0.86
Fe^{2+}	0.24	0.25	0.29	0.31	0.33	0.21	0.12
Mg	0.78	0.62	0.64	0.61	0.94	0.68	1.04
Interlayer							
Ca	0.02	–	0.06	–	0.09	0.04	0.11
Na	0.03	–	0.53	–	0.06	–	0.04
K	0.92	0.74	0.44	0.67	0.58	0.84	0.42
H	1.91	3.02	2.67	3.02	5.15	3.04	6.24
Σ oct.	2.02	2.08	2.00	2.13	2.19	2.04	2.23
Σ alk.	0.97	0.74	1.03	0.67	0.73	0.88	0.57
Charge tet.	–	0.10	0.14	0.13	0.12	0.12	0.21
Charge oct.	0.96	0.63	0.93	0.53	0.70	0.77	0.47

Sources for analyses:

1. Wells (1937, p.102)
2. Levi (1914)
3. Lacroix (1916)
4. Koenig (1916)
5. Heddle (1879)
6. Maegdefrau and Hofmann (1937)
7. Scherillo (1938)

Occurrence:

from vesicular basalt

amygdules in basalt
lavas in Vesuvius
amygdular basalt, pseudomorphs after olivine

CELADONITE

8	9	10	11	12	13	14	15
56.47	52.53	52.58	54.38	56.02	49.85	49.05	49.78
0.13	0.25	0.15	0.14	0.43	–	0.21	0.36
9.09	4.97	6.77	5.41	17.83	4.83	18.17	16.42
12.36	18.62	20.07	14.22	1.14	20.39	6.42	9.74
2.19	4.58	3.33	3.56	2.79	2.49	2.56	3.77
5.98	5.35	6.22	6.40	5.21	1.24	–	–
0.12	0.01	trace	0.25	0.03	4.40	3.10	4.71
1.13	0.58	0.91	0.42	0.68	1.45	1.03	0.75
0.86	0.00	0.05	0.05	–	3.34	0.23	0.15
6.49	7.93	3.33	9.23	9.17	5.80	6.62	7.72
–	–	–	0.15	–	–	–	–
5.32	4.31	6.75	4.80	5.03	3.60	4.56	3.08
	1.15		1.16	1.51	3.23	8.91	3.60
–	–	–	n.d.	–	–	–	–
–	–	–	–	0.12	–	–	0.28
100.14	100.64	100.16	100.17	99.96	100.62	99.86	100.36
3.88	3.78	3.73	3.90	3.88	3.67	3.61	3.51
0.12	0.22	0.27	0.10	0.12	0.33	0.39	0.49
0.62	0.20	0.30	0.36	1.33	0.09	1.19	0.87
0.64	1.01	1.07	0.77	0.01	1.13	0.36	0.51
0.13	0.28	0.20	0.21	0.02	0.15	0.16	0.22
0.61	0.57	0.66	0.68[1]	0.54	0.48	0.34	0.50
0.08	0.04	0.07	0.03	0.05	0.11	0.08	0.05
0.11	–	0.01	0.01	–	0.48	0.03	0.02
0.57	0.73	0.30	0.84	0.81	0.54	0.53	0.70
2.44+	2.07+	3.19	2.30+	2.32	1.77	2.24	1.98
2.00	2.06	2.23	2.02	1.90	1.85	2.05	2.10
0.76	0.77	0.38	0.88	0.86	1.13	0.64	0.77
0.12	0.22	0.27	0.10	0.12	0.33	0.39	0.49
0.74	0.67	0.17	0.83	0.86	1.08	0.35	0.42

8. Scherillo (1938)
9. Bayramgil et al. (1952)
10. Malkova (1956)
11. Lazarenko (1956)
12. Wise and Eugster (1964)
13. Kardymowica (1960)
14. Kvalvaser (1953)
15. Smulikowski (1936)

amygdular basalt, pseudomorphs after olivine

veins and vugs in metamorphic rocks
basalt cavities
amygdule fillings in basalt
in tuffite
veins in oxykeratophyre

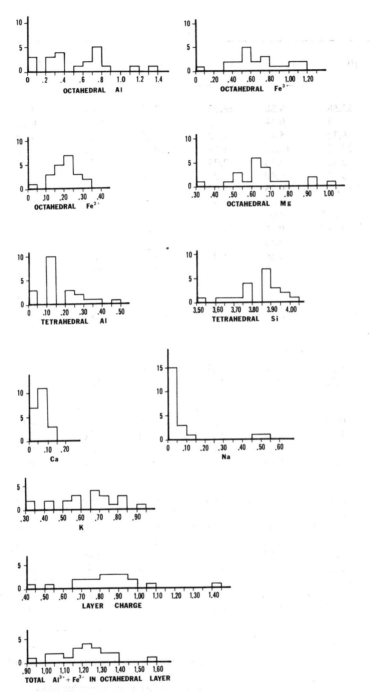

Fig. 10. Histograms showing the distribution of the cations of twenty-one celadonite structural formulas.

The ferric/ferrous iron ratios are similar to those for glauconites. The compositions of the octahedral sheets of celadonite and glauconite are similar. Foster (1969) presented conclusions about the chemical similarities and differences between celadonite and glauconite. It was shown that the range of values for Fe^{2+} in celadonites having less than 0.66 K per unit half cell was lower than in celadonites having greater than 0.65 K per unit half cell. These lower values for Fe^{2+} in the octahedral sheet were accompanied by higher values for the amount of Fe^{3+}, a decrease in bivalent cations, and a lower charge deficiency in the octahedral sheet. The octahedral sheets of celadonite contain on the average considerably more Mg than glauconite (0.63 versus 0.39 or 6.12% MgO versus 3.58% MgO). Fe^{2+} (0.21 versus 0.20) and Al (0.49 versus 0.44) values are similar but octahedral Fe^{3+} (0.72 versus 1.01) averages less for celadonite than for glauconite. The Fe/Mg ratio in celadonite (1.5) is less than half that of glauconites (3.1). This may be due to the decrease in the electronegativity of the oxygen caused by the larger amount of Al substituting for Si in the tetrahedral sheet of glauconite than in celadonite. A similar relation exists between illite and montmorillonite.

The average formula based on the 15 analyses in Table XXIV is:

$$K_{0.64}Na_{0.13}Ca_{0.06}$$
$$Al_{0.49}Fe^{3+}_{0.72}Fe^{2+}_{0.21}Mg_{0.63})(Al_{0.19}Si_{3.81})O_{10}(OH)_2$$

Calculated layer charge is 0.88 with nearly 80% of this charge originating in the octahedral sheet compared with 55% for glauconite.

Total Al + Fe^{3+} occupancy of the three octahedral positions is, in general, less than for the other 2:1 clays. Trivalent occupancy values range from 0.92 to 1.55 with most of the values being less than 1.30, which is the minimum value for the other 2:1 clays including glauconites. The lack of any strong predominance of trivalent over divalent (or divalent over trivalent) ions in the octahedral sheet of celadonite may be the reason that all three octahedral sites are of equal size (Zvyagin, 1957) and may be the best basis of separating celadonites from glauconites. In the other dioctahedral clays the vacant site is larger than the two filled sites.

Radoslovich (1963b) has suggested that celadonite will accommodate a relatively smaller number of trivalent cations in the octahedral sheet than other dioctahedral clays because of the relatively high content of Fe^{3+} compared to Al. The larger Fe^{3+} ions in the octahedral sheet can accommodate the anion-anion repulsion easier than the smaller Al ions can. However, a plot of octahedral Al + Fe^{3+} versus Fe^{3+} of celadonite and glauconite shows only a very vague tendency (if any) for a negative correlation. The Al/Fe^{3+} occupation of the octahedral sheet of celadonite ranges from 0.08/1.13 to 1.33/0.01 and the graphs in Fig.10 indicate there is a continuous series.

Celadonite, with a low trivalent ion octahedral occupancy, differs from the other clays in having, on an average, twice as much Mg. A plot of octahedral Mg versus Al/(Al+Fe^{3+}) (Fig.11) shows that as the relative percentage of Al in the octahedral position decreases, the percentage of octahedral Mg increases. Thus, as the relatively

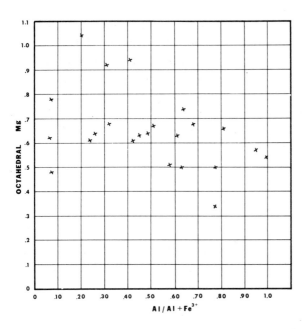

Fig. 11. The relation between octahedral Mg and $Al/(Al + Fe^{3+})$ for twenty-three celadonites.

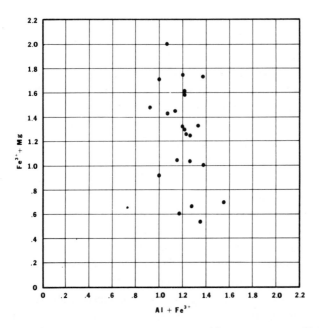

Fig. 12. The relation between octahedral Fe^{3+} + Mg and $Al + Fe^{3+}$ for twenty-two celadonites.

small Al ion is replaced by the larger Fe^{3+} ion, there is an accompanying increase in the even larger Mg ion. Presumably the large Mg ions are necessary to allow for the large shared octahedral edges in celadonites. It would seem that the pure $AlFe^{3+}$ octahedral sheet is not likely to be any more stable than the AlMg (1:1) octahedral sheet. Fig.12 indicates that as the total trivalent population of the octahedral sheet of celadonite decreases and a 50/50 ratio of trivalent and divalent cations is approached, the total amount of large octahedral cations (Fe^{3+} and Mg) increases rapidly.

When an octahedral sheet has appreciably more Fe^{3+} (i.e., glauconite) than is present in celadonite, less Mg is needed to provide the total number of large cations (Fe^{3+}+Mg). This results in an increase in the total trivalent ions in the octahedral positions. In glauconites the Al + Fe^{3+} total is larger than 1.3, dominating the octahedral sheet, and a plot indicates there is no relation between the Mg content and the Al/(Al+Fe^{3+}) ratio.

In general, when either Al or Fe^{3+} is the dominant (greater than 1.0) cation in the octahedral sheet of a 2:1 dioctahedral clay, the maximum Mg content the sheet can accommodate is 0.50–0.60 (0.5 if Fe^{3+} is dominant and 0.6 if Al is dominant). When the Mg content is larger than 0.6, as for most celadonites, seldom is any other cation present in amounts greater than 1.0. This suggests structural control of composition.

The Fe^{2+} content of celadonites (continental origin) and glauconites (marine origin) is identical suggesting that its abundance is not controlled by environmental conditions. Structural control is more likely. Apparently, layer strain is less and the structure more stable when there are 0.20 large Fe^{2+} ions in the octahedral sheet.

Pirani (1963) and Kautz (1965) studied clay minerals that would be described as celadonite because of their occurrence in association with igneous rocks. Kautz plotted two "celadonites" on the muscovite-celadonite-pyrophyllite ternary diagram of Yoder and Eugster (1955) and showed that the "celadonites" were closer to the region of glauconite. Although Pirani agrees with the differences (that have been presented in this paper) between glauconite and celadonite, he feels that they should be considered part of a series with a third member called pholidoides by Smulikowski (1954).

There is sufficient overlap of the composition of celadonite and glauconite to indicate that there is a continuous series and that a genetic classification, while useful, is not definitive.

Chapter 5

SMECTITE

Dioctahedral smectite

Montmorillonite

There are several hundred analyses of montmorillonite in the literature. Most of these montmorillonites were formed by the alteration of volcanic material and most have a relatively small amount of Al substituting for Si in the tetrahedral sheet. One hundred and one analyses were used for the statistical calculations. An additional hundred could have been used, but it is unlikely that they would have significantly changed the statistics.

Only analyses indicating less than 2.10 octahedral cations per $O_{10}(OH)_2$ were used in the calculations. Dioctahedral smectite minerals with an octahedral occupancy larger than 2.10 may exist, but careful analyses seldom show values this large. It appears that most values larger than 2.05 are due to the presence of impurities or to the assignment of interlayer cations to the octahedral position (Foster, 1951). The 101 samples include only those that are in the montmorillonite-beidellite series. The differentiation between the two end members is based on the amount of tetrahedral aluminum. For some time, it was thought that all beidellites were mixtures; however, recent analyses (Greene-Kelly, 1955; Sawhney and Jackson, 1958; Weir and Greene-Kelly, 1962) leave little doubt as to the existence of a high aluminum member of the series.

Tables XXV and XXVI contain average values and other statistical parameters for these 101 samples. Averages for the interlayer cations are of little value as many of the analyses did not include these and, in other instances, pretreatment removed portions of the original cations.

The analyses were taken largely from the following papers: Ross and Hendricks (1945), Kerr et al. (1950), Osthaus (1955), Sawhney and Jackson (1958), Grim and Kulbicki (1961), and a number of single analyses of samples from the United States, Russia and Japan.

Most of these clays were formed by the alteration of volcanic material and basic igneous rocks. A few analyses of gouge clay and hydrothermal alteration deposits are included. Volcanic-derived montmorillonites have been formed in marine and lacustrine environments and in a variety of weathering and terrestial environments. A

TABLE XXV

Statistical data on compositions of 101 montmorillonites–beidellites

Variable		Mean	SE	Range	Standard deviation	SE	Skewness	Kurtosis
No.	name							
1	SiO_2	59.488	0.340	51.20–65.00	3.266	0.262	−0.866**	0.376
2	Al_2O_3	21.934	0.333	15.20–34.00	3.192	0.361	1.295**	2.700**
3	Fe_2O_3	3.770	0.273	0.00–13.61	2.618	0.250	1.045**	1.366**
4	FeO	0.197	0.043	0.00– 1.61	0.416	0.131	5.203**	34.688**
5	MgO	3.548	0.170	0.09– 7.38	1.630	0.106	0.180	−0.448
6	CaO	1.176	0.128	0.00– 4.23	1.227	0.067	0.771**	−0.894
7	Na_2O	0.824	0.112	0.00– 3.74	1.070	0.110	1.709**	1.904**
8	K_2O	0.342	0.049	0.00– 1.82	0.471	0.071	2.282**	6.380**
9	TiO_2	0.250	0.040	0.00– 2.90	0.384	0.103	4.023**	24.442**
10	H_2O^+	8.380	0.193	5.21–13.75	1.852	0.145	0.807**	0.262

SE − skewness = 0.251; SE − kurtosis = 0.498; **significant at 0.01 level.

TABLE XXVI

Statistical data on structural fomulas of 101 montmorillonites-beidellites

Variable		Mean	SE	Range	Standard deviation	SE	Skewness	Kurtosis
No.	name							
1	Oct. Al	1.492	0.016	1.10–2.00	0.158	0.014	0.393	1.250**
2	Oct. Fe^{3+}	0.187	0.013	0.00–0.68	0.132	0.012	1.117**	1.443**
3	Oct. Fe^{2+}	0.007	0.002	0.00–0.09	0.016	0.003	3.539**	14.736**
4	Oct. Mg	0.354	0.015	0.01–0.71	0.153	0.010	0.053	−0.400
5	Tet. Al	0.158	0.013	0.00–0.66	0.127	0.013	1.397**	2.387**
6	Tet. Si	3.837	0.014	3.34–4.00	0.137	0.015	−1.467**	2.843**
7	Int. Ca	0.007	0.005	0.00–0.49	0.050	0.024	9.408**	91.685**
8	Int. Na	0.010	0.005	0.00–0.52	0.051	0.014	5.357**	29.275**
9	Int. K	0.004	0.002	0.00–0.16	0.022	0.007	5.711**	33.645**

SE − skewness = 0.240; SE − kurtosis = 0.476; **significant at 0.01 level.

selection of montmorillonite-beidellite analyses and structural formulas are given in Tables XXVII and XXVIII.

The significance of average values depends on how representative the samples are of the total population, the heterogeneity of the population, errors of sampling and errors of analysis (Shaw, 1961). With the present set of data it is impossible to assay the magnitude of these various sources of variation. Shaw states that "... the apparent variability or variance of a set of analyses will always be greater than the variability of the material sampled. The total variance is biased by the errors picked up between the outcrop and the analysis statement". This is particularly true when the analyses are tabulated from the literature. Although Ahrens (1954) used histograms to indicate that trace element distributions follow a log-normal law, Shaw (1961), using cumulative frequency-distribution plots on normal probability paper, indicated that in some instances, the distribution was closer to being arithmatic normal. Ahrens found normal distributions for elements with small dispersion but believed the normal distribution was not significant because of the small dispersion. The calculated skewness values indicate that all of the distributions except octahedral Al and Mg, and MgO are significantly skewed. The skewness is positive except for Si and H_2O.

Fig.13 and Fig.14 contain histograms showing the distribution of the oxides and the cations in the various structural positions. Tetrahedral Al ranges from 0.00 to 0.66 and averages 0.16. The histogram suggests a log-normal distribution. The mode is at about 0.10. Approximately two-thirds of the samples have less than 0.20 of the four tetrahedral positions filled with Al. It should be noted that tetrahedral Al in the illites has a more normal distribution. The histograms of these two clays show relatively little overlap; however, this is in part due to the small number of beidellite analyses in the literature. There are only a few samples with more than 0.40 tethrahedral Al. The histogram and cumulative curve for tetrahedral Al suggest that the distribution is bimodal. The major mode is at approximately 0.10 and a secondary mode at about 0.30.

Octahedral Al averages 1.49 (1.53 for illites) and ranges from 1.10 to 2.00. The distribution is normal and the mode is nearly identical to the mean. Ninety-two percent of the values are within the relatively narrow range of 1.25—1.75% Al. Octahedral Fe^{3+} appears to have log-normal distribution with a modal value of 0.15, slightly less than the mean (0.19). Octahedral Fe^{2+} has a similar distribution although in most samples it was either not present or not determined. Octahedral Mg has a near-normal distribution with the mode approximately the same as the mean (0.354). Both octahedral Al and Mg are normally distributed in the illites as well as in the montmorillonites. In both minerals the Al distribution is more restricted (leptokurtic) than the Mg. The distribution pattern would indicate that the amount of Al in the octahedral position is relatively fixed. Most of the variation is due to the Mg and Fe^{3+} and Fe^{2+} with the latter two showing the widest variation and, therefore, likely being the least significant in defining the mineral species.

TABLE XXVII

Chemical analyses and structural formulas of some montmorillonites of various origins

	1	2	3	4	5	6	7	8
SiO_2	53.98	51.14	51.95	48.24	50.53	48.60	51.7	51.
Al_2O_3	15.97	19.76	18.02	20.30	19.31	18.40	20.2	22.
Fe_2O_3	0.95	0.85	0.21	2.29	7.25	1.21	0.77	–
FeO	0.19	–	–	0.11	–	0.07	–	–
MgO	4.47	3.22	5.10	2.06	2.60	1.88	3.4	3.
CaO	2.30	1.62	2.09	2.10	0.72	2.25	2.5	3.
Na_2O	0.13	0.04	0.04	0.36	0.41	0.35	0.09	
K_2O	0.12	0.11		0.05	0.34	0.28	0.84	0.
TiO_2	0.08	–	0.02	0.08	0.75	Tr	–	
H_2O^+	9.12	7.99	7.60	7.20	7.90	8.44	7.4	17.
H_2O^-	13.06	14.81	15.60	17.94	10.66	17.64	13.0	
Total	100.62[1]	99.75[2]	100.65[3]	100.75[4]	100.64[5]	99.47[6]	99.99[7]	99.

	1	2	3	4	5	6	7	8
Octahedral								
Al	1.48	1.64	1.43	1.65	1.40	1.68	1.58	1.6
Fe^{3+}	0.05	0.05	0.01	0.24	0.40	0.08	0.05	–
Fe^{2+}	–	–	–	0.01	–	0.01	–	–
Mg	0.52	0.36	0.58	0.24	0.28	0.23	0.37	0.3
	2.05	2.05	2.02	2.14	2.08	2.00	2.00	2.0
Tetrahedral								
Al	0.00	0.12	0.15	0.22	0.28	0.06	0.14	0.2
Si	4.00	3.88	3.85	3.78	3.72	3.94	3.26	3.7
Interlayer								
Ca/2	0.39	0.20	0.32	0.34		0.38	0.41	
K	–	–	–	–		0.03	–	
Na	0.02	0.02	0.01	0.01		0.06	0.02	
Mg/2	–	–	–	–		–	0.08	
C.E.C. (mequiv./100 g)	122							
Layer charge	0.37	0.33	0.67	0.05	0.32	0.30	0.51	0.3

[1] MnO = 0.06%; [2] CuO ≠ 0.01%; [3] Mn = 0.02%; [4] P_2O_5 = 0.02%; [5] MnO = 0.02%, P_2O_5 = 0.15%; [6] P_2O_5 = 0.05%; [7] SO_4 = 0.09%; [8] P_2O_5 = 1.42%, MnO = 0.18%; [9] F = 0.39%; [10] MnO = 0.01%.

1. Kerr et al. (1950): altered rhyolitic and andesitic tuff, Santa Rita, New Mex., U.S.A.; analyst Ledoux & Co.
2. Ross and Hendricks (1945): nests penetrating a shale, Montmorillon, France; analyst R.C. Wells.
3. Blokh and Sidorenka (1960): cavity filling inextrusive rocks, Selongin Daura, U.S.S.R. (called nefedyevite).
4. Yoshikawa (personal communication, 1960): beds in Tertiary tuff, Tottori, Japan.
5. Ross and Hendricks (1945): Upper Cretaceous bentonite, Booneville, Miss., U.S.A.; analyst J.G. Fairchild.
6. Yoshikawa (personal communication, 1960): vein cutting granite, Tottori, Japan.
7. Fournier (1965): hydrothermally altered plagioclase, Ely, Nev., U.S.A.; analysts Elmore, Botts, Chloe, Artis and S

	10	11	12	13	14	15	16	17
7.38	48.30	59.32	46.95	59.57	49.90	47.0	50.01	50.72
1.27	16.98	18.54	27.26	18.35	19.17	15.4	19.36	18.12
0.66	4.04	2.86	2.26	2.76	2.24	12.32	2.36	2.41
	–	–	0.32	0.11	0.01	–	0.70	1.02
0.42	3.74	3.22	1.39	3.78	0.01	–	4.40	4.29
0.78	2.90	0.21	–	0.22	3.22	1.36	1.20	0.80
0.12	0.30	1.34	0.20	2.20	0.47	0.24	1.09	3.00
0.08	1.25	0.6	0.36	1.05	1.34	0.87	0.35	0.62
	0.51	0.08	0.00	0.12	0.12	0.42	0.20	0.24
9.08	8.76	7.45	10.55	7.42	0.21	0.50	6.28	6.87
9.60	12.80	6.39	11.10	6.78	15.69	7.2	14.59	11.90
9.39	99.97[9]	100.01	100.43[10]	102.36	100.09	97.71	100.54	99.99

	10	11	12	13	14	15	16	17
45	1.33	1.53	1.84	1.46	1.58	1.10	1.50	1.46
60	0.24	0.15	0.12	0.14	0.13	0.68	0.13	0.13
	–	–	0.02	0.01	–	–	0.04	0.06
05	0.43	0.33	0.15	0.38	0.37	0.20	0.44	0.45
10	2.00	2.01	2.13	1.99	2.08	1.98	2.11	2.10
44	0.23	0.00	0.54	0.00	0.16	0.07	0.22	0.15
56	3.77	4.05	3.46	4.01	3.84	3.93	3.78	3.85
13	0.49	0.03		0.03	0.08		0.08	0.03
01	0.12	0.05		0.09	0.01		0.01	0.02
02	0.05	0.18		0.29	0.20		0.15	0.42
						90	100	
19	0.66	0.30	0.32	0.42	0.29	0.31	0.37	0.36

. Ross and Hendricks (1945): pegmatitic clay, Branchville, Conn., U.S.A.; analyst G.J. Brush.
. Oyawoye and Hirst (1964): hydrothermal vein in granite, Ropp, northern Nigeria; analyst R. Lambert.
. Weaver (unpublished): detrital in Cretaceous shale, Denver Basin, Colo., U.S.A.; analyst E.H. Oslund.
. Whitehouse and McCarter (1958): Upton, Wyo., U.S.A. montmorillonite in sea water 54-60 mo.
. Ross and Hendricks (1945): clay replacing calcareous shells, Pontotoc, Miss., U.S.A.; analyst J.G. Fairchild.
. Whitehouse and McCarter, (1958): soil, Houston, Texas, U.S.A.
. Whitehouse and McCarter, (1958): soil, San Saba, Texas, U.S.A.
. Sawhney and Jackson (1958): soil, Aina Haina, Hawaii, derived from basic rocks and alluvium.
. Alietti and Alietti (1962): Lower Miocene bentonitic marl, Gemmano, Italy.
. Alietti and Alietti (1962): Lower Miocene bentonitic marl, Gemmano, Italy.

TABLE XXVIII

Beidellite analyses and structural formulas

	1	2	3	4		1	2	3	4
					Octahedral				
SiO_2	49.01	55.80	59.30	44.02	Al	1.34	1.85	1.98	1.8
Al_2O_3	20.5	28.60	36.11	28.90	Fe^{3+}	0.39	0.02	0.02	0.2
Fe_2O_3	6.85	0.41	0.50	5.15	Fe^{2+}	–	–	–	0.0
FeO	–	–	–	0.33	Mg	0.28	0.20	0.01	0.0
MgO	2.08	2.03	0.10	0.50	Σ	2.01	2.07	2.01	2.1
CaO	0.17	2.23	0.02	0.68					
Na_2O	0.58	2.09	3.98	0.17	Tetrahedral				
K_2O	0.95	0.48	0.11	0.50	Al	0.27	0.35	0.52	0.7
TiO_2	0.17	0.26	–	0.53	Si	3.73	3.65	3.48	3.2
H_2O^+	12.1	9.70							
H_2O^-	10.5	–		18.64	Interlayer				
					Ca/2		0.31		
	103.0	99.60	100.12*	99.42	K		0.04		
					Na		0.01		
					C.E.C.				
					(mequiv./100 g)	115		130**	
					Layer charge	0.52	0.34	0.50	0.3

1. Sawhney and Jackson (1958): Houston soil, less than 0.08 microns.
2. Heystek (1962): Castle Mountain, Ivanpah, Calif., U.S.A., hydrothermally altered.
3. Weir and Green-Kelley (1962): Black Jack Mine, Beidell, Colo., U.S.A., Gouge clay.
4. Ross and Hendricks (1945): Ancon, Canal Zone, laminated organic clay.
*Loss on ignition; 6.3% (ignited weight).
**Mequiv./100 g ignited weight).

Calculated layer charge (Fig.14) has a well-developed mode occurring between the values 0.30–0.35 (average 0.41). Forty-five percent of the values are in this narrow range. The minimum charge is close to 0.30, legitimate maximum values are as large as 0.60–0.65.

Table XXIX contains the correlation coefficients for the ions in the various structural positions. Only the obvious correlations have significant values and these are not too high. There is a negative correlation between octahedral Al and octahedral Fe^{3+} and Mg and an obvious high negative correlation between tetrahedral Si and Al.

Grim and Kulbicki (1961) have suggested that there are two types of montmorillonite, which differ primarily in the amount of octahedral Mg (Table XXX). The Cheto type has approximately one-fourth of the Al replaced by Mg (average of 0.55 octahedral positions). It is suggested that the Mg ions are regularly distributed in a hexagonal arrangement. The Wyoming-type montmorillonites have considerably less Mg (average 0.19 octahedral positions). There is no significant difference in the tetrahedral makeup of the two types. Most of the measurements Grim and Kulbicki made indicate that these two types of montmorillonite are significantly different. Many of the samples they studied were intermediate in composition and were believed to be mixtures of the two end member types. Size fractionation of one sample indicated that the coarser fraction exhibited the characteristics (when heated) of the Cheto- type and the finer fraction of the characteristics of the Wyoming-type.

The histogram of octahedral Mg (Fig.14) indicates a symmetrical distribution for Mg and does not suggest a bimodal distribution; however, a probability plot suggests the possibility of a secondary mode in the low Mg values. The frequency distribution would indicate that if these two end member types exist, most montmorillonites are mixtures of the two.

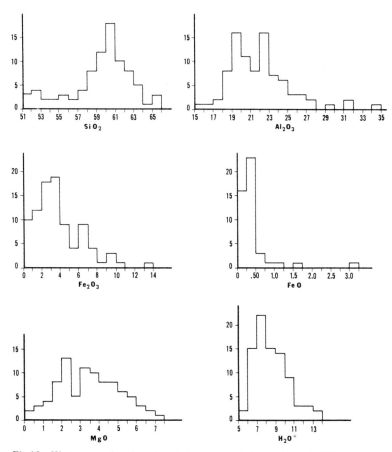

Fig.13. Histograms showing the distribution of the oxides of 101 montmorillonite-beidellite analyses.

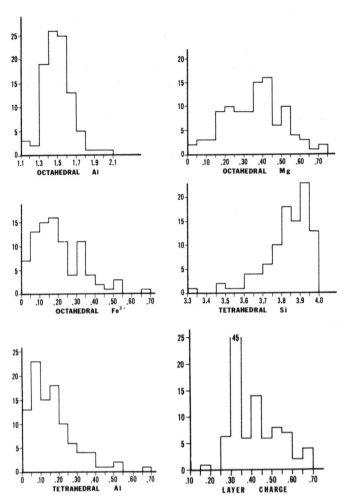

Fig. 14. Histograms showing the distribution of the cations of 101 montmorillonite-beidellite structural formulas.

SiO_2 has a negatively skewed distribution with a mean slightly smaller than the well-developed mode (60–61%). Both Al_2O_3 and MgO appear to have a bimodal distribution although that of MgO probably is not real. The Al_2O_3 distribution is more suggestive of two end-member types of montmorillonite (Grim and Kulbicki, 1961) than any of the other distributions. Fe_2O_3 and FeO and H_2O^+ values appear to have a log-normal distribution. The distribution of the ferrous and ferric oxides reflects the prevalence of low values.

Correlation coefficients in Table XXXI indicate that Fe_2O_3 and H_2O^+ are negatively related to SiO_2. The same relation holds for glauconites (Table XVIII); Fe_2O_3 and SiO_2 are inversely related in illites. If high SiO_2 indicates low tetrahedral

TABLE XXIX

Correlation coefficients for the cations in 101 montmorillonites-beidellites

Variable		1	2	3	4	5	6
No.	name						
1	Oct. Al	1.000					
2	Oct. Fe^{3+}	−0.564	1.000				
3	Oct. Fe^{2+}	0.057	−0.021	1.000			
4	Oct. Mg	−0.539	−0.346	−0.156	1.000		
5	Tet. Al	0.336	0.144	0.049	−0.387	1.000	
6	Tet. Si	−0.277	−0.134	−0.029	0.322	−0.927	1.000

charge, then as the layer charge shifts from the tetrahedral to the octahedral sheet, the amount of Fe^{3+} in the octahedral sheet decreases; however, correlation coefficients of the structural formula do not suggest that this is the case. MgO is positively correlated with SiO_2 and negatively correlated with Al_2O_3 and Fe_2O_3 which reflects an increase in octahedral charge as the tetrahedral charge decreases. The negative relation between SiO_2 and H_2O^+ may indicate that as the tetrahedral charge increases, interlayer water is more tightly bonded (or trapped) and is not all released as H_2O^-

Ross and Hendricks (1945) redefined beidellite as the aluminum-rich end member of the dioctahedral montmorillonites. Many of their samples were later found to be mixtures and for some time the concept of a high-aluminum montmorillonite was in considerable disrepute. Recently, Weir and Greene-Kelly (1962) made a careful analysis of purified material from the Black Jack Mine from Beidell, Colorado, and definitely established that it is monomineralic and an ideal Al-rich end member (Table XXVIII). They suggest that "beidellites and montmorillonites should be divided at the composition at which the lattice charges from octahedral and tetrahedral substitution equal one another". The layer charge for beidellite does not appear to be any larger than for montmorillonites.

There is a complete isomorphous series from $Si_{4.0}Al_{0.0}$ to $Si_{3.50}Al_{0.50}$. Values larger than $Al_{0.50}$ have been reported (Ross and Hendricks, 1945; Sawhney and Jackson, 1958) but in these samples, the sum of the octahedral cations is excessive, usually being on the order of 2.2, and gives a positive charge to the octahedral sheet. Whether such high octahedral occupancy is real, or not, has not been satisfactorily determined.

Plots of total Al versus tetrahedral Al and total Al versus octahedral Al (Fig.15) both show a good linear relation. The slopes of the two plots indicate that as the total amount of Al increases, the absolute amount in the octahedral sheet increases at a faster rate than the absolute amount in the tetrahedral sheet. On a relative basis as the

TABLE XXX

Chemical analyses of montmorillonites (After Grim and Kulbicki, 1961)

	Cheto-type						
	1	2	3	4	5	6[+]	7
SiO_2	61.77	62.23	63.07	61.55	60.90	60.80	65.9
Al_2O_2	19.85	21.03	18.46	20.44	20.71	22.15	19.1
TiO_2	0.24		0.28				
Fe_2O_3	1.95*	1.75	1.99*	2.02	2.06	0.07	1.0
FeO		0.48		0.38	0.36		0.7
MgO	5.56	5.70	7.38	6.06	6.84	4.44	4.5
CaO	1.89	0.00	0.24	0.00	0.30	3.74	0.0
Na_2O	0.07	0.65	0.16	0.30	0.23		0.1
K_2O	0.09	0.00	0.16	0.00	0.00		0.0
H_2O^+	7.72	7.38	7.17	8.29	8.02	8.71	7.2
Total	99.14	99.22	98.91	99.04	99.42	99.91	98.8
H_2O^-	9.49	8.41	13.03	7.67	7.08	14.75	5.4
Al Tetrahedr.	0.09	0.14	0.07	0.16	0.19	0.02	
Si Tet.	3.91	3.86	3.93	3.84	3.81	3.98	++
Al Octahedr.	1.38	1.39	1.28	1.34	1.33	1.67	1.4
Fe^{3+} Oct.	0.09	0.08	0.09	0.09	0.09		0.0
Fe^{2+} Oct.		0.02		0.02	0.02		0.0
Mg Oct.	0.54	0.55	0.71	0.58	0.65	0.38	0.4
Σ Oct.	2.01	2.04	2.08	2.03	2.09	2.05	2.0
Layer charge	60	59	54	67	59	25	4 (avg.)

Analyses based on dry (105°C) weight of samples;
* = total Fe as Fe_2O_3; [+] = crude clay; ++ = contains free silica.

Location of samples:

Cheto-type montmorillonites
1. Cheto, Ariz., U.S.A.
2. Otay, Calif., U.S.A.
3. Burrera, Jachal, San Juan, Argentina.
4. El Retamito, Retamito, San Juan, Argentina.
5. Mario Don Fernando, Retamito, San Juan, Argentina.
6. Tatatilla, Vera Cruz, Mexico.
7. Itoigawa, Niigata, Japan.

Wyoming-type

	9	10	11	12	13	14	15	16
4.80	62.00	62.30	62.70	59.73	60.22	60.76	59.91	58.67
4.54	23.42	23.50	22.20	24.30	23.67	23.08	21.97	27.34
					0.34	0.38	0.33	
1.27	3.74	3.35	4.62	5.54	6.28*	6.10*	6.72*	3.64
0.56	0.32	0.37	0.48	0.37				0.38
1.60	0.93	1.95	2.00	2.10	1.46	1.44	2.15	2.00
0.00	0.68	0.31	0.58	0.00	0.13	0.17	0.34	0.00
0.40	0.72	0.40	0.01	0.80	0.09	0.13	0.09	0.62
0.60	2.63	0.03	0.12	0.22	0.19	0.21	0.11	0.18
5.71	5.21	6.45	7.06	6.59	6.86	6.07	6.66	7.04
0.48	99.65	98.66	99.77	99.65	99.24	98.34	98.28	99.87
5.22	6.44	7.81	7.13	13.70	6.81	7.65	8.81	10.88
0.04	0.08	0.10	0.08	0.24	0.20	0.16	0.17	0.34
3.96	3.92	3.90	3.92	3.76	3.80	3.84	3.83	3.66
1.72	1.66	1.64	1.55	1.56	1.56	1.56	1.48	1.67
0.06	0.18	0.15	0.22	0.26	0.30	0.29	0.32	0.17
0.03	0.02	0.02	0.03	0.02				0.01
0.15	0.09	0.19	0.19	0.20	0.14	0.14	0.21	0.19
1.96	1.95	2.00	1.99	2.04	2.00	1.99	2.01	2.05
34	34	31	33	34	34	30	35	42 (avg. 34)

Wyoming-type montmorillonites

8. Hojun Mine, Gumma, Japan.
9. Tala, Heras, Mendoza, Argentina.
10. Crook County, Wyo., U.S.A.
11. Rokkaku, Yamagata, Japan.
12. Amory, Miss., U.S.A.
13. Santa Elena, Potrerillos, Mendoza, Argentina.
14. San Gabriel, Potrerillos, Mendoza, Argentina.
15. Emilia, Calingasta, San Juan, Argentina.
16. Sin Procedencia, Argentina.

TABLE XXXI

Correlation coefficients for the oxides of 101 montmorillonites-beidellites

Variable No. name	1	2	3	4	5	6	7	8	9	10
1 SiO_2	1.000									
2 Al_2O_3	−0.342	1.000								
3 Fe_2O_3	−0.520	−0.259	1.000							
4 FeO	0.205	−0.197	−0.037	1.000						
5 MgO	0.416	−0.532	−0.407	−0.019	1.000					
6 CaO	−0.172	−0.058	−0.231	0.110	0.101	1.000				
7 Na_2O	0.091	−0.034	−0.037	−0.075	−0.012	−0.329	1.000			
8 K_2O	−0.274	−0.153	0.328	0.004	−0.193	−0.113	0.042	1.000		
9 TiO_2	−0.414	−0.151	0.386	0.024	−0.131	−0.045	−0.053	0.231	1.000	
10 H_2O^+	−0.640	−0.079	0.271	−0.206	−0.085	0.214	−0.333	0.226	0.369	1.000

total Al increases the proportion of tetrahedral Al increases and that of octahedral Al decreases.

As the histograms of both montmorillonite and illite indicate, a minimum of 1.30 octahedral positions must be filled with Al to afford a 2:1 sheet structure. The relative consistency of the relationship shown by the two plots suggest that the distribution of the Al between the two structural positions is an equilibrium distribution. If the line representing the tetrahedral Al versus total Al is extended to intersect a total Al value of 3, approximately 0.8 of the Al would be in the tetrahedral position. This appears to be the maximum value found for illites. (If the octahedral graph is extended, 2.40 octahedral positions would have to be filled to accommodate a total of 3 Al.) The data indicate that few montmorillonites have a total Al value larger than 2.00; however, some beidellites with values as large as 2.25 Al fall on trend. Similar plots for illites show a wider dispersion but the trends appear to approximately parallel the montmorillonite trends (Fig.15). Most of the illites have a total Al value larger than 2.0 and none exceed 2.70. The discontinuity between the illite and montmorillonite data or between the less than 2.0 Al and greater than 2.0 Al samples suggest that there is no continuous isomorphous series between illite and montmorillonite. When the total Al is larger than 2.0, a much higher proportion of it occurs in the tetrahedral sheet and correspondingly less in the octahedral sheet. This could suggest that some octahedral Al has moved to the tetrahedral position during metamorphism; however, the shift in position is accompanied by an increase in total Al. Once the threshold value of 2.0–2.1 Al is exceeded, additional Al is divided between tetrahedral and octahedral sheets in the approximate proportion of 70% tetrahedral and 30% octahedral. This could suggest significant migration between sheets, but probably indicates that when montmorillonite converts to illite, Al from an external source is added and due to higher temperature conditions (~200°C), a relatively large proportion assumes tetrahedral coordination as compared to the low-temperature montmorillonites.

In muscovite, potassium plays a major role in controlling the size of the tetrahedral and octahedral sheets (Radoslovich, 1963a). The interlayer cations in montmorillonite are of little value in helping ease the strain caused by misfit between the two types of sheets and adjustment must be made by isomorphous substitution. The graphs in Fig.15 indicate that when there is no Al substitution in the tetrahedral sheet, the minimum amount of Al in the octahedral sheet is on the order of 1.25, the remaining 0.75 positions being occupied by the larger Fe and Mg ions. The layer charge has a maximum value of 0.75. Both the size of the two sheets and the layer charge allow the development of a layer structure. As the total amount of Al is increased the addition to the tetrahedral sheet increases its size and the addition to the octahedral sheet causes a decrease in size. There is also a decrease in layer charge as well as a more equal distribution of the charge between the two sheets. When the total Al is as high as 2.00, the strain, due to the misfit between the sheets, is too large to be accommodated without the help of interlayer potassium to assist in causing tetrahedral rotation (Weaver, 1967). The substitution of additional Al in the tetrahedral sheet is required in order to increase the layer charge sufficiently to fix K.

When the total amount of Al in the structure is enough to fill more than two structural positions, the amount of octahedral Al decreases by approximately 0.25. These positions are filled by the larger Mg (0.65 Å ionic radius) and Fe (Fe^{2+} 0.75 Å; Fe^{3+} 0.60 Å) ions. Fe becomes relatively more important than Mg in these high Al clays. The average octahedral Fe content in illite is 0.26 and in montmorillonite 0.19. In the montmorillonites, the tendency is for the higher Al samples to have the high Fe content. Ramberg (1952) has shown that when Al is substituted for Si in the silicate structure, the electronegativity of the oxygen decreases and the Fe/Mg ratio is likely to increase, all other conditions being equal. This relation appears to hold in a general way in the 2:1 clays (Fe/Mg = 0.54, tetrahedral Al = 0.16, montmorillonite; Fe/Mg = 0.93, tetrahedral Al = 0.60, illite).

Ross and Hendricks (1945) and others have noted that Mg is essential for the formation of montmorillonite. Mg increases the size and the charge of the octahedral sheet and tends to decrease layer strain. However, its major contribution may be in enabling Al to take six-fold coordination under basic conditions. At low temperature Al tends to take six-fold coordination in an acid environment and four-fold in a basic environment (DeKimpe et al., 1961). Mg takes only the six-fold coordination. In the basic environments in which montmorillonite and illite form, it is likely that Al tends to resist six-fold coordination and Mg may be necessary to nucleate the octahedral coordination.

In a series of hydrothermal experiments, Hawkins and Roy (1963) found 2–3%

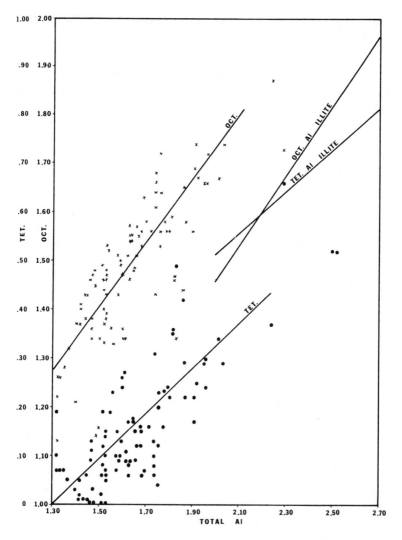

Fig.15. The relation of total Al to octahedral Al and tetrahedral Al of 101 montmorillonite-beidellites and 24 illites.

MgO was necessary to form montmorillonite from a variety of starting materials. When Mg was not present, analcite was formed and all the Al was presumably tetrahedrally coordinated. When CaO was added to the starting material, rather than MgO, tobermorite was formed and Al did not enter into any crystalline silicate structure. Although Hawkins and Roy (1963) and Keller (1964) believe the role of Mg is to precipitate the Si, it is possible that its major contribution is, in some manner, assisting the Al in assuming six-fold coordination under conditions where it would normally take four-fold coordination.

A source of error in chemical analyses of montmorillonites (and in other clays) that is not commonly checked is the presence of amorphous material, particularly Si and Al. Table XXXII lists structural formulas given by Osthaus (1955) for montmorillonites which were purified by size fraction and by extraction with 0.5 N NaOH to remove amorphous Si and Al. In six analyses dissolved silica ranged from 3.6 to 8.4% and alumina from 0.6 to 2.25%. Amorphous silicon dioxide should be expected in most montmorillonites derived from volcanic material. The source glass has more Si than is required for the 2:1 layer and the excess must be leached from the glass. Much of the Si is deposited in the sediments underlying the bentonite bed in the form of chert but it is to be expected that the extraction would not be complete and a portion of the colloidal Si would remain in the bentonite bed.

The cation exchange capacity of montmorillonites ranges from 70 to 130 mequiv./ 100 g. Most of the exchange capacity (80%) is due to substitution within the structure, but a lesser amount (20%) is due to the charges at the edge of the sheets (Hendricks, 1942). Na, Ca, Mg and H are the most common interlayer cations found in nature. Table XXXIII from Osthaus (1955) shows the type exchangeable cations from a selection of montmorillonites. Table XXXIV illustrates the range in composition occurring in the Black Hills Cretaceous bentonite (Knechtel and Patterson, 1962). These two tables indicate that although the variety of exchange cations is limited, the cation

TABLE XXXII

Structural formulas of montmorillonites before and after extraction of easily soluble Si and Al (After Osthaus, 1955)

Before	After
Belle Fourche, S.D., U.S.A.	
$(Al_{1.56}Fe_{0.20}Mg_{0.23})(Si_{3.86}Al_{0.14})O_{10}OH_2$	$(Al_{1.53}Fe_{0.23}Mg_{0.21})(Si_{3.77}Al_{0.23})O_{10}OH_2$
Clay Spur, Wyo., U.S.A.	
$(Al_{1.53}Fe_{0.20}Mg_{0.25})(Si_{3.88}Al_{0.12})O_{10}OH_2$	$(Al_{1.52}Fe_{0.26}Mg_{0.27})(Si_{3.83}Al_{0.17})O_{10}OH_2$
Plymouth, Utah, U.S.A.	
$(Al_{1.40}Fe_{0.24}Mg_{0.38})(Si_{3.82}Al_{0.18})O_{10}OH_2$	$(Al_{1.34}Fe_{0.27}Mg_{0.43})(Si_{3.73}Al_{0.27})O_{10}OH_2$
Chambers, Ariz., U.S.A.	
$(Al_{1.40}Fe_{0.17}Mg_{0.43})(Si_{3.88}Al_{0.12})O_{10}OH_2$	$(Al_{1.37}Fe_{0.19}Mg_{0.46})(Si_{3.84}Al_{0.16})O_{10}OH_2$
Polkville, Miss., U.S.A.	
$(Al_{1.45}Fe_{0.88}Mg_{0.47})(Si_{4.00})O_{10}OH_2$	$(Al_{1.39}Fe_{0.10}Mg_{0.58})(Si_{3.94}Al_{0.06})O_{10}OH_2$
Otay, Calif., U.S.A.	
$(Al_{1.35}Fe_{0.05}Mg_{0.67})(Si_{3.98}Al_{0.02})O_{10}OH_2$	$(Al_{1.32}Fe_{0.06}Mg_{0.71})(Si_{3.95}Al_{0.05})O_{10}OH_2$

TABLE XXXIII

Base-exchange determinations on untreated montmorillonite clays (After Osthaus, 1955)

Location	Cations (mequiv./100 g)					Anions (mequiv./100 g)			Base-exchange capacity cations-anions	Base-exchange capacity NH_4^+ dist.
	Na^+	K^+	Ca^{2+}	Mg^{2+}	H^+	SO^{2-}	Cl^-	CO^{2-}		
Merritt, British Columbia	23.88	1.49	32.50	10.00			7.80	3.50	56.57	57.5
Santa Rita, New Mexico	2.26	1.91	62.9	29.0					96.07	97.6
Belle Fourche, South Dakota	84.80	1.70	7.80	8.50		18.30	5.70		78.86	79.8
Little Rock, Arkansas	1.29	1.27	2.14	22.50	48.70				75.90	78.8
Polkville, Mississippi	2.00	1.00	95.30	21.00			1.60		117.70	113.7
Amory, Mississippi	1.09	2.76	70.00	28.00		1.70		29.00	71.15	71.4
Otay, California	40.70	1.27	20.00	70.40		1.50	8.40		122.47	122.5
Plymouth, Utah	0.45	1.36	99.00	39.00				33.00	106.80	111.8
Clay Spur, Wyoming	82.60	1.74	12.45	8.80		13.00	3.60	10.50	78.49	75.5
Chambers, Arizona	5.50	1.20	100.00	20.00		1.00	4.40	2.00	119.30	117.8

TABLE XXXIV

Cation exchange data (in mequiv./g) for samples of bentonite from the Northern Black Hills district, Montana, Wyoming, and South Dakota (After Knechtel and Patterson, 1962)

Bentonite	Locality	Ca	Mg	Na	K	H	Total	Determined
A	27	0.53	0.24	0.13	0.02	trace	0.79	0.77
	58a	0.18	0.23	0.55	0.01	–	0.80	0.81
B	30	–	–	0.01	0.01	0.32	0.54	0.69
	46	0.59	0.26	0.07	0.03	–	0.89	0.91
Clay spur	19	0.03	0.09	0.68	0.03	–	0.85	0.84
	26	0.12	0.28	0.36	0.01	0.02	0.79	0.83
	28	–	–	–	–	–	–	–
	33	–	–	–	–	–	–	–
	34	0.31	0.24	0.31	0.02	–	0.87	0.80
	43	–	–	–	–	–	–	–
	56	–	–	–	–	–	–	–
	56	0.03	0.12	0.81	0.02	–	0.91	0.84
	56	–	–	–	–	–	–	–
	62	0.02	0.09	0.68	0.01	–	0.81	0.83
	70	0.26	0.21	0.35	0.01	0.02	0.85	0.89
	76	–	0.14	0.48	0.03	–	0.65	0.64
	76	0.16	0.19	0.49	0.02	–	0.85	0.82
	77	–	–	–	–	–	–	–
		0.20	0.06	0.65	0.01	–	0.78	0.78
		0.01	0.01	0.64	0.02	–	0.68	0.68
		0.20	0.03	0.76	0.02	–	1.00	0.85
		0.29	0.09	0.48	0.01	–	0.87	0.67
		0.31	0.11	0.29	0.02	–	0.73	0.66
D	70	–	–	–	–	–	–	–
F	4	0.67	0.24	0.42	0.02	–	1.35	0.87
	42	0.34	0.41	0.20	0.02	–	0.78	0.85
	54	0.12	0.27	0.42	0.02	–	0.82	0.83
G	38	0.22	0.53	0.42	0.03	–	1.00	0.93
I	78	0.54	0.16	0.18	0.01	0.06	0.95	0.88
	78	0.64	0.44	–	0.02	–	1.06	0.95

Total ion-exchange determinations by Dorothy Carroll; exchangeable-cation analyses by W.W. Brannock.

suites are extremely variable. Although any of the cations, except K, can predominate, few samples tend to be mono-ionic.

Little effort appears to have been made to relate the exchangeable cation suite to the depositional environment. The final suite is dependent upon the composition of the water in the environment of deposition; however, it is also dependent upon the original cation suite, time, type of clay, and post-depositional leaching, among other things.

Keller (1964) reported that Ca is the dominant exchangeable cation on montmorillonite in equilibrium with river water. In sea water Ca and also Na (Carroll and Starkey, 1960) tend to be replaced by Mg which becomes the dominant exchangeable cation. In many ancient clays, Na is the most abundant exchangeable cation; therefore, this abundance of Na appears to be inconsistent with the above data. Recently Hanshaw (1964) conducted exchange experiments with compacted clays and found that the order of cation selectivity is dependent upon whether a clay is dispersed or compacted. He found that compacted montmorillonite preferred cations in the following order: $K^+ > Na^+ > H^+ > Ca^{2+} > Mg^{2+}$. It may be that in dispersed marine montmorillonites Mg is the predominant exchange cation, but as the mud is compacted by burial, Na replaces a portion of the Mg. This has been confirmed by Weaver and Beck (1971a).

Many montmorillonites, particularly the beidellite-type, have units with a sufficiently high layer charge to fix K in the interlayer position and cause the layers to contract to 10 Å. This K is not usually exchangeable using the normal methods of determining exchange capacity. Wyoming montmorillonite when exposed to KOH or sea water loses approximately 15% exchange capacity and a similar percentage of layers contract to 10Å. Beidellite-type clays may fix several percent of K and have most of their layers contracted to 10Å (Weaver, 1958).

The C.E.C. of montmorillonite does not appear to vary with grain size (Osthaus, 1955); however, there is a relation between cation type and flake size. In most montmorillonite clays and shales the Na is concentrated in the finer fraction and the Ca in the coarser fraction (McAtee, 1958). This is probably because Na allows much greater interlayer expansion which leads to the shearing-off of thin flakes.

Roberson et al. (1968) studied three size fractions of some montmorillonites with respect to changes in interlayer charge and the possibility of isomorphous substitution with differences in particle size. They suggested that the physical differences observed among the various size fractions of the montmorillonites are not due to differences in composition or interlayer charge, but to the intermeshing of many small crystals into a micro-aggregate by contemporaneous growth.

Soil scientists (Sawhney, 1958; Rich and Cook, 1963; and others) have established that Al and Fe hydroxides are commonly present in the interlayer position of soil montmorillonites. This material is not readily exchangeable and is probably responsible for a great deal of error in calculated structural formulas.

The calculated layer charge for illites and montmorillonites shows some overlap although the average value for the illites is considerably higher (0.69 versus 0.41). This overlap is largely due, exclusive of analytical errors, to the presence of expanded layers in some of the illites and contractable or potentially contractable layers in some of the montmorillonites. The maximum C.E.C. for montmorillonites is in the range of 130–145 mequiv./100 g. A layer charge of 0.55–0.60 per $O_{10}(OH)_2$ is required to afford such a high C.E.C. A small portion of this charge will be due to unsatisfied edge bonds rather than substitution within the structure. C.E.C. increases as layer charge increases up to a maximum charge of approximately 0.6. If potassium is available, layers with a charge higher than this will fix the potassium and contract to 10 Å. Approximately 7% K_2O is required to satisfy a layer charge of 0.6.

TABLE XXXV

Structural formulas for montmorillonites from Cretaceous bentonites of Black Hills region (Wyoming, South Dakota)

Sample		Formula
1	bluish gray	$(Al^{3+}_{1.60}Fe^{3+}_{0.06}Fe^{2+}_{0.18}Mg^{2+}_{0.22})(Si^{4+}_{3.91}Al^{3+}_{0.09})O_{10}(OH)_2$
	olive green	$(Al^{3+}_{1.61}Fe^{3+}_{0.13}Fe^{2+}_{0.06}Mg^{2+}_{0.20})(Si^{4+}_{3.92}Al^{3+}_{0.08})O_{10}(OH)_2$
2	bluish gray	$(Al^{2+}_{1.62}Fe^{3+}_{0.00}Fe^{2+}_{0.13}Mg^{2+}_{0.20})(Si^{4+}_{3.92}Al^{3+}_{0.08})O_{10}(OH)_2$
	olive green	$(Al^{3+}_{1.62}Fe^{3+}_{0.13}Fe^{2+}_{0.05}Mg^{2+}_{0.19})(Si^{4+}_{3.90}Al^{3+}_{0.10})O_{10}(OH)_2$
3		$(Al_{1.64}Fe^{3+}_{0.15}Fe^{2+}_{0.02}Mg_{0.19})(Si_{3.90}Al_{0.10})O_{10}(OH)_2$
4		$(Al_{1.54}Fe_{0.16}Mg_{0.33})(Si_{3.91}Al_{0.09})O_{10}(OH)_2$
5		$(Al_{1.55}Fe_{0.15}Mg_{0.33})(Si_{3.92}Al_{0.08})O_{10}(OH)_2$
6		$(Al_{1.63}Fe_{0.17}Mg_{0.25})(Si_{3.93}Al_{0.07})O_{10}(OH)_2$
7		$(Al_{1.61}Fe_{0.18}Mg_{0.23})(Si_{3.87}Al_{0.13})O_{10}(OH)_2$
8		$(Al_{1.55}Fe^{3+}_{0.19}Fe^{2+}_{0.02}Mg_{0.26})(Si_{3.88}Al_{0.12})O_{10}(OH)_2$
9		$(Al_{1.56}Fe^{3+}_{0.03}Mg_{0.19})(Si_{3.85}Al_{0.15})O_{10}(OH)_2$

1,2. Knechtel and Patterson (1962): Clay spur bentonite bed, Crook Co., Wyoming; analyst M.D. Foster.
3. Grim and Kulbicki (1961): Crook Co., Wyoming.
4. Kerr et al. (1950): Clay spur, Wyoming; 5. Upton, Wyoming; 6. Bell Fourche, S.D. A portion of the octahedral Mg undoubtedly belongs in the interlayer position.
7. Foster (1954): Belle Fourche, South Dakota.
8. Ross and Hendricks (1945): Upton, Wyoming.
9. Mackenzie (1963): Wyoming bentonite.

The data on mixed-layer illite-montmorillonites indicate that when all the K_2O is assigned to the contracted 10 Å layers, the minimum value is 7% K_2O. Weaver (1964b) has noted that 2M illites containing less than 7% K_2O show peak broadening indicative of the presence of expanded layers and more recently (Weaver and Beck, 1971a), of chloritic layers. Montmorillonites exist which have some layers with a layer charge larger than 0.6 per $O_{10}(OH)_2$ and are expanded but this is due to the absence of potassium. The basic problem is the distribution of charge. Most data suggest the layer charge is not homogeneous and clays are a mixture of high-charged layers (~ 1.0) and lower-charged layers. The seat of the charge would seem to be of minor importance. If much of the charge originates in the octahedral sheet, a value slightly larger than 0.6 might be necessary to cause contraction, perhaps as large as 0.7.

There appears to be some relation between composition and mode of origin but with the present data, it is difficult to deduce. In general, the high-Mg, low-Al montmorillonites are most likely to have formed by the alteration of volcanic material and the high-Al beidellite type, to have been of hydrothermal origin or to have been formed as gouge clay and soil clay. In the soil clays much of the "excess" Al is present as hydroxyl Al in the interlayer position.

The composition of montmorillonites formed from similar volcanic materials in a marine environment can be quite uniform over fairly large areas. Table XXXV contains a number of structural formulas obtained from the Cretaceous bentonite beds of the Black Hills region of Wyoming, Montana, and South Dakota. The first four formulas were obtained from analyses of two partially weathered blocks from one mine. The major difference is due to the oxidation of the iron by weathering (Knechtel and Patterson, 1962). The analyses are surprisingly similar. Much of the small variation that does exist is due to the failure of some of the analysts to distinguish between octahedral and exchangeable Mg. Such a uniform composition could indicate that this is the most stable composition for montmorillonite. More likely, it indicates an ash of uniform composition, altered in a marine environment for a relatively similar time interval. Also it should be pointed out that these are all commercial samples and the variability would perhaps be greater if non-commercial samples were included.

In contrast, a study of the Tertiary bentonite beds of Texas (Roberson, 1964) revealed considerable variation in chemical composition. Although no chemical analyses were made, the ability of the various montmorillonites to contract to 10 Å when treated with K indicates a wide variation in layer charge. These bentonites are believed to have altered in a lacustrine environment and some are believed to be detrital, possibly having been formed in soils.

Thus, the variations in composition appear to be primarily due to environmental differences whereas the consistant composition of Wyoming bentonites is related to the uniformity of the shallow marine environment in which they were deposited. The data indicate that Texas bentonites fix more K than Wyoming bentonites. This likely means that they also have a larger amount of tetrahedral Al. Tetrahedral Al increases as total Al increases and Mg decreases. These suggested differences in composition are in the direction that would be expected for montmorillonites formed in these two different environments. The montmorillonites forming in the bottom of the South Pacific from volcanic materials are reported to be relatively rich in iron (Arrhenius,1963).

Van der Kaaden and Quakernaat (1968) reported a manganiferous smectite in which Mn^{2+} occupies an octahedral position. They gave the following structural formula:

$$(Al_{1.52}Fe^{3+}_{0.06}Fe^{2+}_{0.01}Mg_{0.92}Mn^{2+}_{0.03})(Si_{3.85}Al_{0.15})O_{10}(OH)_2 0.40M^+.$$

Nontronite

Nontronite is the iron-rich dioctahedral smectite. The amount of octahedral iron in the dioctahedral smectites varies from 0.0 to 2.0. Frequency plots do not suggest any bimodal distribution. The modal value for octahedral iron in montmorillonite is approximately 0.15 (Fe^{3+}). Frequency values then drop rapidly and there is a continuous even distribution of values between 0.5 and 2.0 (Fe^{3+}); however, these data only include nineteen analyses classed as nontronite and of these, five have values of 1.9–2.0. It is possible that if more samples were analyzed, a well-developed, high-iron mode would be found. Table XXXVI lists typical chemical analyses of nontronites. Tetrahedral Al ranges from 0.23 to 0.75, which is essentially the same as for the beidellites. Like the beidellites, when tetrahedral Al is larger than 0.50, the sum of the octahedral cations is appreciably greater than 2.00. Octahedral Mg ranges from 0.01 to 0.58 though the average value is only 0.14. Apparently when there is an abundance of iron available, Mg is not necessary for the formation of montmorillonite. The excess octahedral population (in both nontronites and beidellites) might suggest the presence of some interlayered trioctahedral sheets; however, if this were the case, the samples with the larger octahedral populations should have a relatively high Mg content (rather than Al and Fe^{3+}). As the opposite seems to be the case it is likely that some hydroxy-Al,-Fe are present in the interlayer position.

Although all of the iron is usually assigned to the octahedral sheet, leaching experiments by Osthaus (1954) and Mössbauer studies by Weaver et al. (1967) indicate that much of it may be in the tetrahedral sheet. Table XXXVII shows the structural formula for the Garfield, Washington, nontronite as usually calculated and the formula calculated by Osthaus. There is no strong correlation between Mg and Fe; however, as Ross and Hendricks (1945) indicated, the samples with the highest Fe content tend to have a lower Mg content. This same weak relation exists in the montmorillonites-beidellites.

The total amount of divalent cations in the octahedral sheet is considerably less than in the montmorillonites, and most octahedral sheets presumably have a net positive charge. In this respect the nontronites resemble the beidellites. Most of the

TABLE XXXVI

Chemical analyses and structural formulas of some iron-bearing dioctahedral montmorillonites and nontronites

	1	2	3	4	5	6	7
SiO_2	56.91	39.52	43.05		41.88	43.51	39.92
Al_2O_3	22.65	18.48	6.40		11.90	2.94	5.37
Fe_2O_3	6.29	12.60	17.86		26.20	28.62	29.46
FeO	0.11	–	0.10		–	0.99	0.28
MgO	3.62	3.27	4.46		0.10	0.05	0.93
CaO	1.47	2.72	2.92		0.67	2.22	2.46
Na_2O	0.18	0.10	–		–	–	–
K_2O	0.74	0.50	–		0.52	–	–
TiO_2	0.65	1.88	–		–	–	0.08
H_2O^+	7.34	12.10	23.93		7.67	6.62	7.00
H_2O^-		9.04	23.93		10.78	14.05	14.38
	99.96	100.21	100.04		99.72	100.02	99.88

1. Altschuler et al. (1963): marine clay associated with Bone Valley phosphates, Fla., U.S.A.; analyzed by H. Kramer.
2. Alietti (1960): hydrothermal alteration of basaltic tuff.
3. Koster (1960): altered basalt; Cr_2O_3 0.87%.
4. Sawhney and Jackson (1958): soil developed on olivine basalt, Waipiate, New Zealand; analyzed by L.D. Swindale.

negative charge originates in the tetrahedral sheet. If octahedral populations are as indicated, the net layer charge is generally smaller, 0.25–0.35, than for the low-iron montmorillonites.

The analyses of dioctahedral smectites indicate that octahedral Fe and Al form a continuous isomorphous series and any subdivisions must be arbitrary. As the octahedral Al/Fe ratio increases, there is a tendency for octahedral Mg to increase (although

TABLE XXXVII

Structural formulas for Garfield nontronite

$(Al_{0.05}Fe_{1.93}Mg_{0.12})(Al_{0.50}Si_{3.50})O_{10}(OH)_2$

$(Al_{0.48}Fe_{1.49}Mg_{0.09})(Al_{0.09}Fe_{0.38}Si_{3.52})O_{10}(OH)_2$

	1	2	3	4	5	6	7
Octahedral							
Al	1.45	1.00	0.44	0.47	0.51	0.06	0.03
Fe^{3+}	0.31	0.72	1.18	1.37	1.58	1.86	2.02
Fe^{2+}	0.01	–	0.01	–	–	0.07	–
Mg	0.35	0.34	0.58	0.27	0.01	0.01	–
Σ	2.12	2.06	2.21	2.11	2.10	2.00	2.05
Tetrahedral							
Al	0.29	0.72	0.22	0.62	0.63	0.24	0.50
Si	3.71	3.28	3.78	3.38	3.37	3.76	3.50
Interlayer							
Ca/2	0.21	Ca/2	0.38				0.35
K	0.06	Mg/2					–
Na	0.02						0.02
Layer charge	29	88	18	56	34	32	35

5. Ross and Hendricks (1945): #66 alteration zone in gneiss, Spruce Pine, N.C., U.S.A.; analyzed by L.T. Richardson.
6. Ross and Hendricks (1945): #56 veinlets in garnet-pyroxene rock, Woody, Calif., U.S.A.; analyzed by E.V. Shannon.
7. Kerr et al. (1950): alteration of basalt, Manito, Wash., U.S.A.; analyzed by W.C. Bowden.

the pure Al end member exists). There is no apparent relation between the octahedral Al/Fe ratio and the amount of tetrahedral Al.

Nontronite apparently forms under the same general environmental conditions as the low-iron montmorillonites. It is formed by hydrothermal alteration and as vein fillings. It is commonly formed by both the hydrothermal alteration and surface weathering of basalt. Nontronite is the dominant clay in some soils (Ross and Hendricks, 1945). Arrhenius (1963) found that much of the authigenic montmorillonite in the pelagic muds of the Pacific Ocean has a relatively high iron content.

Trioctahedral smectite

The trioctahedral smectites are quite variable in composition, particularly in the octahedral sheet. The Mg-rich end member which contains little Al in either the octahedral or tetrahedral sheet has been called stevensite (Faust and Murata, 1953;

TABLE XXXVIII

Chemical analyses and structural formulas of stevensite and hectorite samples

	1	2	3	4			1	2	3	4
SiO_2	57.30	53.37	55.02	55.17	Octahedral					
Al_2O_3	none	0.91	1.12	0.33		Al	–	0.06	0.08	0.02
Fe_2O_3	0.32	0.66	–	0.12		Fe^{3+}	0.02	0.04	–	–
FeO	none	–	0.70	trace		Fe^{2+}	–	–	0.05	–
MnO	0.21	0.09	–	trace		Mg	2.88	2.81	2.57	2.65
MgO	27.47	25.29	24.89	24.51		Mn	0.02	–	–	–
CaO	0.97	–	0.54	0.90		Li	–	0.04	0.10	0.33
Na_2O	0.03	0.28	0.94	2.20		Σ	2.92	2.95	2.80	3.00
K_2O	0.03	0.02	0.43	0.08	Tetrahedral					
Li_2O	–	0.13	0.36	1.14		Al	0.00	0.02	0.02	0.00
TiO_2	–	0.09	0.08	0.01		Si	4.00	3.98	3.98	4.00
CO_2	–	–	0.30	0.63	Interlayer					
F	none	0.91	3.22	4.75	cations					
H_2O^+	7.17	8.76	6.42	2.84		Ca/2	0.15		0.02	0.02
H_2O^-	6.69	9.94	7.66	8.93		Mg/2			0.02	–
Total	100.19	100.38	101.68	101.87*		Na			0.13	0.28
						K			0.04	0.01
				C.E.C.						
				(mequiv./						
				100 g)		36.00		75.1	115.00
				Layer						
				charge		0.14	0.10	0.44	0.33

*inclusive 0.21 Cl, 0.05 P_2O_5.
1. Faust and Murata (1953): hydrothermal, pseudomorphous after pectolite, Springfield, N.J., U.S.A.; analyst K.J. Murata.
2. Bradley and Fahey (1962): lacustine, Green River Formation, Wyom., U.S.A.; analyst J.J. Fahey.
3. Faust et al. (1959): authigenic lake deposite associated with marl, eastern Morocco; analyst J.J. Fahey.
4. Ames et al. (1958): hot springs alteration of zeolite in alkaline lake, Hector, Calif., U.S.A.; analyst S.S. Goldrich.

Table XXXVIII). Brindley (1955) has suggested that stevensite is a mixed-layer talc-saponite; however, Faust et al. (1959) considered it to be a defect structure with a random distribution of vacant sites in the octahedral sheets. A small proportion of domains with few or no vacancies would then be present having characteristics of talc. The layer charge in stevensite is due to an incompletely filled octahedral sheet (Faust and Murata, 1953). This deficiency is minor (0.05–0.10) and the resulting cation exchange capacity is only about one-third that of the dioctahedral montmorillonites (100 mequiv./100 g.).

Most stevensite and mixtures of stevensite and other minerals (commonly talc and pectolite) have a hydrothermal origin, being derived from pectolite by the action of magnesium-bearing solutions. Bradley and Fahey (1962) described an authigenic occurrence of stevensite in the lacustrine Green River Formation of Wyoming which presumably formed under strongly saline conditions.

Hectorite is similar to stevensite in having little or no tetrahedral substitution; however, the octahedral sheet has a significant Li content (Table XXXVIII). The hectorite from Hector, California, contains 0.33 octahedral Li. A sample described by Faust et al. (1959) contains only 0.10 Li and a sample described by Bradley and Fahey (1962) contains 0.04 octahedral Li. There could presumably exist a continuous range, although the upper limit is not known. Values as high as 1.45 are reported for the trioctahedral micas (Radoslovich, 1962). Layer charge is due both to Li substitution and cation deficiencies in the octahedral sheet. Appreciable F^- is present proxying for OH^-.

The material from the Hector area of California is believed to have formed by the action of hot spring waters containing Li and F on clinoptiolite. The Mg was obtained from the alkaline lake waters (Ames and Goldich, 1958). The material from Morocco is associated with marls and is believed to be authigenic. These two types of trioctahedral smectite appear to be the only ones with a relatively pure Si tetrahedral sheet. No analyses were found which indicated tetrahedral Al values between 0.02 and 0.30. Analyses of saponite indicate there is complete isomorphous substitution between the range $Si_{3.70} Al_{0.30}$ and $Si_{3.08} Al_{0.92}$ (Table XXXIX). Caillère and Hénin (1951) reported an analysis of a fibrous expanded clay (diabantite) which had a tetrahedral composition of $Si_{3.17} Al_{0.49} Fe^{3+}_{0.34}$. There is some question as to whether this should be classified as a smectite; regardless, it indicates the possibility of Fe^{3+} substitution in the tetrahedral sheets of the trioctahedral 2:1 clays.

For saponites there is a good linear relationship between the amount of tetrahedral Al and the total amount of octahedral positions filled by trivalent ions (Fig. 16). Iron is by far the most abundant trivalent cation. Octahedral Al ranges from 0.00 to 0.27 with most samples having less than 0.05 positions occupied by Al. Fe^{3+} values range from 0.00 to 0.52. Total Fe ranges from 0.00 to 1.45. The Fe^{3+}/Fe^{2+} ratio is variable and probably not significant because of variations due to weathering. Sudo (1954) described an authigenic clay associated with iron sand beds that has 1.45 of the total of the 3.01

TABLE XXXIX

Chemical analyses and structural formulas of some saponites

	1	2	3	4	5	6	7
SiO_2	50.01	43.62	44.00	40.46	43.98	39.68	40.44
Al_2O_3	3.89	5.50	10.60	6.30	10.15	3.93	14.52
Fe_2O_3	0.21	0.66	trace	3.56	7.85	19.82	1.85
FeO	–	–	trace	4.89	5.32	1.12	2.37
MnO	–	0.06	–	0.24	0.32	0.19	0.15
MgO	25.61	24.32	24.30	20.71	18.02	11.21	21.11
CaO	1.31	2.85	2.00	1.94	2.78	2.37	2.32
Na_2O	–	0.08	–	0.25	–	–	0.42
K_2O	–	0.04	–	0.32	–	–	trace
TiO_2	0.04	0.00	–	–	0.16	0.37	–
CO_2	–	–	–	–	–	–	–
H_2O^+	12.02	5.48	6.20	4.24	9.24	6.16	6.31
H_2O^-	7.28	17.42	12.60	13.33	6.24	15.11	10.65
Total	100.37	100.03	99.70	100.23*	100.39	99.96	100.14

1. Cahoon (1954): hydrothermal alteration of dolomitic limestone, Milford, Utah, U.S.A.; analyst W. Savournin.
2. Mackenzie (1957): in vesicles in basalt, Allt Ribhein, Skye; analyst J.B. Craig.
3. Ross and Hendricks (1945): from mine, San Bernadino, Calif., U.S.A.; analyst W.L. Gibson.

positions occupied by Fe^{2+} (Table XXXIX). An Al-rich saponite described by Mongiorgi and Morandi (1970) is an Al-saponite with 0.38 octahedral Al and only 0.11 octahedral Fe^{3+}.

The upper limit for the number of R^{3+} cations in octahedral positions is essentially 0.50 or approximately 85% occupancy by R^{2+} cations. This compares with a minimum R^{2+} occupancy value of about 75% for trioctahedral micas (Radoslovich, 1963b) and 65% for dioctahedral smectites and illites. The composition of the octahedral sheet of the trioctahedral smectites falls within the compositional limits established for the trioctahedral micas (Foster, 1960). All samples fall within the

	1	2	3	4	5	6	7
Octahedral							
Al	0.04	0.02	0.27	0.11	0.00	0.04	0.38
Fe^{3+}	0.01	0.04	–	0.21	0.40	0.00	0.11
Fe^{2+}	–	–	–	0.32	0.35	1.45	0.15
Mg	2.85	2.91	2.73	2.35	2.09	1.52	2.40
Mn	–	–	–	0.02	0.02	–	0.01
Σ	2.90	2.97	3.00	2.91	2.91	3.01	3.05
Tetrahedral							
Al	0.30	0.50	0.67	0.83	0.58	0.38	0.92
Si	3.70	3.50	3.33	3.17	3.42	3.62	3.08
Interlayer							
Ca/2							
Mg/2							
Na							
K							
C.E.C. (mequiv./100 g)							76
Layer charge	0.45	0.50	0.67	0.49	0.46	0.32	0.44

*includes P_2O_5 0.14%.
4. Konta and Sindelar (1955): fissure fillings of amphibolites, Caslov, Czechoslovakia.
5. Miyamoto (1957): altered basalt.
6. Sudo (1954): authigenic clay matrix in tertiary iron sand bed, Moniwa, Japan.
7. Mongiori and Morandi (1970): hydrothermal alteration of poligenic breccia, Rossewa, Italy.

phlogopite–Mg-biotite field. Octahedral substitution is much more restricted in the trioctahedral smectites than in the micas.

The Mg-octahedral sheet is larger than the tetrahedral sheet (b-axis) and is forced to contract (Radoslovich, 1963b). Substitution of the larger Fe^{2+} would presumably increase the strain beyond tolerable limits. Such adjustments would be less readily expected in minerals formed at low temperatures. Although substitution of Al and Fe^{3+} would tend to decrease the size of the octahedral sheet, the net positive charge of the sheet is increased (0.10–0.50), decreasing the anion-anion repulsion. Vacancies in the octahedral sheet generally increase as octahedral R^{3+} increases, but this seldom

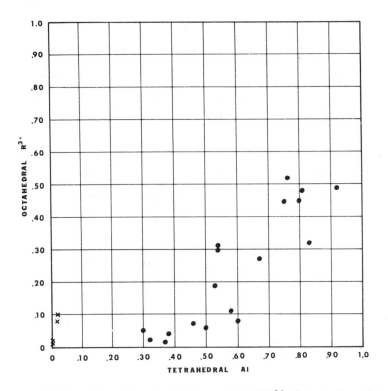

Fig.16. The relation of tetrahedral Al to octahedral R^{3+} of twenty-two trioctahedral montmorillonites. x = stevensite and hectorite; ● = saponite.

results in a negative charge for the layer. The number of R^{3+} cations the octahedral sheet can accommodate is probably controlled by the amount of Al^{3+} substitution in the tetrahedral sheet (Fig.16).

The maximum amount of Al^{3+} tetrahedral substitution that 2:1 clays minerals formed at low temperatures can accommodate appears to be 0.80–0.90 per four tetrahedra. While this appears to place an upper limit on the amount of R^{3+} octahedral substitution, it is not clear why the limit should be such a low value. The dioctahedral smectites can accommodate more substitution (R^{2+} for R^{3+}) in the octahedral sheet than can the dioctahedral micas. The reverse situation exists for trioctahedral equivalents. In the latter clays octahedral R^{3+} increases as tetrahedral Al increases. Thus, as one sheet increases its negative charge, the other tends to increase its positive charge. This is likely to introduce additional constraints on the structure. In the dioctahedral clays substitution in either sheet affords them a negative charge and substitution in one sheet is not predicted by substitution in the other sheet; thus, one might expect more flexibility.

The strain in saponites produces a lath morphology. These laths vary considerably in width; however, the data are insufficient to relate lath width to composition. There is little data on cation exchange capacity but, on the basis of calculated layer charge, it should be as high or higher than that for dioctahedral smectites.

Most of the saponites described have a hydrothermal origin and have usually been formed by the action of either hydrothermal or surface waters on basic rocks. The iron-rich saponite described by Sudo (1954) occurs as a clay matrix in a Tertiary iron sand bed and is presumably authigenic. A metamorphic origin is proposed for saponite from metalimestones (Wilson et al., 1968).

All of the saponites that have been discussed were grown from solution or were found by the alteration of non-micaceous minerals or glasses. Similar type clays can form by the weathering of biotite and phlogopite. MacEwan (1954) has described such a clay (cardenite) from Scottish soils. The formula he gives is:

$$(Al_{0.44}Fe^{3+}_{0.68}Fe^{2+}_{0.15}Mg_{1.48}Ca_{0.12}) (Si_{3.06}Al_{0.90})O_{10}(OH)_2$$

On the basis of this formula both the octahedral and tetrahedral sheets have vacancies. It is unlikely that this structural formula accurately describes a single mineral species. Serdyuchenko (1953) has described similar weathered clays from the Jurassic of Russia:

$$(Na_{0.02}K_{0.32}Ca_{0.03}$$
$$(Mg_{0.85}Fe^{2+}_{0.10}Ni_{0.01}Al_{0.76}Ti_{0.04}Fe^{3+}_{0.66}Cr_{0.03}) (Si_{3.38}Al_{0.62})O_{10}(OH)_2$$

A great many of the less common cations occur in the octahedral sheet of the trioctahedral smectites. Zn, Cr, Ni, Cu, Ti, Mn, and V have all been reported as occurring in significant amounts. The first five can occur as predominant ions. The zinc-bearing smectite, sauconite, is more common and occurs as a purer clay than any of these others. A number of these clays have been described by Ross (1946) and Faust (1951). Analyses of sauconite are reported by Ross (1946); some of these data are reproduced in Table XL along with two other analyses of impure samples. Ross summarized:

"In the formulas of sauconite (Nos. 1–7, 11 and 12) the Al in tetrahedral positions ranges from about 0.50 to 1, and in most of them it lies between 0.60 and 0.80. In formula No.1 the tetrahedral group ($Al_{0.99} Si_{3.01}$) is like that in muscovite and is high for a mineral of the montmorillonite group. Most of the sauconite formulas contain more Al than some of the montmorillonites, in which Al may approach 0 while Si may become approximately 4. Zinc is the dominant ion in octahedral positions in all the formulas calculated and ranges from 1.50 to nearly 2.90, varying reciprocally with Al and Fe and to a lesser extent with Mg. The total number of ions in octahedral positions, indicated by Σ ranges from 2.70 to 3.06."

These clays have considerably more octahedral Al than any of the other trioctahedral smectites. Those with high octahedral Al (0.73–0.79) have a low octahedral population (2.70–2.80). These latter values are the lowest that have been reported for the clay minerals deposited from solution. A minimum of approximately 60% of the

TABLE XL

Chemical analyses and structural formulas of some sauconites (samples 1-5 after Ross, 1960)

	1	2	3	4	5	6	7
SiO_2	34.46	35.95	37.10	33.59	38.70	29.70	28.44
Al_2O_3	16.95	6.57	14.18	6.01	16.29	17.40	30.80
Fe_2O_3	6.21	2.36	0.30	0.28	3.91	2.49	–
MgO	1.11	1.26	1.10	0.70	1.62	1.23	0.00
ZnO	23.10	33.70	28.19	39.33	22.48	29.74	11.12
CuO	–	–	0.02	0.10	–	–	1.87
MnO	–	0.04	0.02	0.12	0.06	–	–
CaO	–	0.62	1.22	1.90	trace	3.60	0.80
Na_2O	–	0.44	0.24	0.13	0.43	–	–
K_2O	0.49	0.10	0.13	0.07	0.32	–	–
TiO_2	0.24	0.07	trace	0.03	0.30	–	–
H_2O^-	6.72	11.34	8.82	10.68	7.50	15.48	16.50
H_2O^+	10.67	7.24	8.90	6.98	8.38		9.32
Total	99.95	99.69	100.22	99.92	99.99	99.64	98.85

1. Friedensville, Pa., U.S.A.; analyst J.G. Fairchild.
2. New Discovery Mine, Leadville; analyst J.G. Fairchild.
3. Yankee Doodle Mine, Leadville, analyst J.G. Fairchild.
4. Coon Hollow Mine, Ark., U.S.A.; analyst M.K. Carren.
5. Plattesville district, Wis., U.S.A.; analyst S.K. Cress.

occupied octahedral sites contain divalent cations as compared to a minimum value of 85% for saponites. Excess positive octahedral charge is high, ranging from 0.18 to 0.76. In all Ross's formulas the negative tetrahedral sheet is larger than the positive sheet, but in No.7 the difference is only 0.04. This would seem to be an unlikely situation. There is the distinct possibility that some of the Zn may not be present in the clay layers. Sauconite has been formed hydrothermally and by weathering of zinc-rich hydrothermal deposits.

	1	2	3	4	5	6	7*
Octahedral							
Al	0.74	0.22	0.79	0.04	0.73	0.77	1.80
Fe^{3+}	0.40	0.17	0.02	0.02	0.43	0.19	
Mg	0.14	0.18	0.14	0.10	0.41	0.18	
Zn	1.48	2.40	1.35	2.89	1.23	1.58	0.74
Cu							0.12
Mn				0.01			
Σ	2.76	2.97	2.30	3.06	2.80	2.72	2.66
Tetrahedral							
Al	0.99	0.53	0.70	0.65	0.80	0.21	0.26
Si	3.01	3.47	3.30	3.35	3.20	3.79	3.74
Interlayer							
Ca/2		0.13	0.23	0.40		0.29	0.08
Na		0.09	0.05	0.02			
K				0.01			
Layer charge	33	20	129	49	4	19$^+$	36$^+$

*PbO = 0.72%

6. Hydrothermal deposits, Kugitong Range: Kuliev (1959).
7. Altyn-Tojkan: EmKeev (1958).

Serdyuchenko (1933), McConnell (1954), and Ross (1960) have reported analyses of Cr-rich clays ranges from 1 to 15% (Table XLI). Some of the clays are formed by ground water alteration of serpentine and probably volcanic material. It is questionable if any of the samples are pure. McConnell's (1954) material has a lath-shaped morphology. Chukhrov and Anosov (1950) described a copper-bearing (20.96%) montmorillonite from the oxidized zone of weathered sulfide ores (Table XLI). Spangenberg (1938) described a nickel-rich montmorillonite.

TABLE XLI

Chemical analyses of some trioctahedral smectites rich in Cr_2O_3 and CuO

	1	2	3
SiO_2	43.14	44.47	43.88
Al_2O_3	16.07	19.63	13.25
Fe_2O_3	4.70	10.85	–
Cr_2O_3	5.02	1.02	–
CuO	–	–	20.96
FeO	0.61	0.47	–
MnO	0.21	trace	–
NiO	0.30	0.10	–
MgO	3.45	1.86	–
CaO	1.46	1.50	–
TiO_2	0.96	0.20	–
H_2O^+	7.32	20.00	7.52
H_2O^-	16.66		7.02
Others	0.16	0.29	–
Total	100.06	100.39	92.63

1. Serdyuchenko (1933): below ore bed in the Cheghet-Lakhair Valley.
2. Serdyuchenko (1933): emerald mines, Ural Mountains.
3. Chukhrov and Anosov (1950): in oxidized layer of weathered sulfide ores.

Chapter 6

CHLORITE

Macroscopic chlorite

The fundamental unit layer of chlorite consists of a 2:1 layer plus a hydroxide sheet. The layer thickness is 14 Å. Chlorites are a group of hydrous silicates of magnesium, aluminum and iron in widely varying proportions. Most are trioctahedral, but dioctahedral and mixed dioctahedral-trioctahedral types exist. Because of the wide range of ionic substitutions, a great many names have been given to various members of the chlorite group. The structural determinations, chemical analyses, and classification of chlorites are based largely on studies of macroscopic chlorites. Chlorite is abundant as a clay-sized mineral, but most of it is derived from macroscopic chlorite; thus, much of the data on the coarser chlorites are directly applicable to the clay chlorites.

The chemistry of the chlorites has been reviewed by Hey (1954), Foster (1962) and Deer et al. (1962). Hey and Foster have presented classification schemes. All chlorites have replacement of Si by Al which affords the tetrahedral sheets a net negative charge. This charge is balanced by the substitution of Al and Fe^{3+} for Mg and Fe^{2+} in the two octahedral sheets.

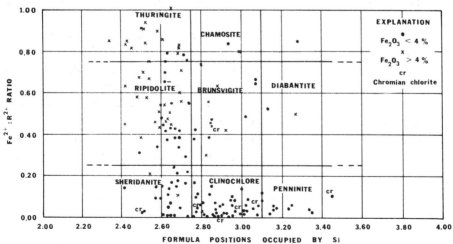

Fig.17. Classification of chlorites based on the two principal types of ionic replacement. (After Foster, 1962.)

Foster (1962) calculated the structural formulas for 150 selected chlorite analyses. These formulas indicate that the Si content ranges from 2.34 to 3.45 per four tetrahedral positions. Most samples fall in the 2.40–3.20 range (Fig.17), the distribution being highly skewed towards the higher Si values. Most chlorites tend to have a much higher tetrahedral Al content than 2:1 clays. (Some of the 1:1 trioctahedral clays are the only clay minerals with tetrahedral Al contents as high as that of most chlorites.)

The tetrahedral charge is balanced by substitution of Al and Fe^{3+} in the hydroxide sheet and in the octahedral sheet in the 2:1 layer. Foster assumed that the R^{3+} cations were evenly distributed between the two sheets, but recent structure analyses indicate that this is not necessarily true. Foster found:

"...that octahedral Al is not often closely equivalent to tetrahedral Al, being higher in about one-fourth of the formulas, and lower in about one-half. If lower, some other octahedral trivalent cation, usually Fe^{3+}, is present and proxies for Al in providing sufficient octahedral positive charge to balance the tetrahedral negative charge. Octahedral occupancy is close to 6.00 only if the sum of the octahedral trivalent cations is approximately equal to tetrahedral Al. If it is greater, as is most usual, octahedral occupancy is less than 6.00 formula positions by an amount equal to about one-half the excess of octahedral trivalent cations over tetrahedral Al, indicating that the excess octahedral trivalent cations replace bivalent cations in the ratio of 2:3."

Octahedral occupancy ranges from 5.46 to 6.05. R^{3+} octahedral occupancy ranges from approximately 10 to 47% of the filled positions with most of the values ranging from 15 to 35%. In general, these values are similar to those for the other clay minerals, although the larger values are larger than those found in any of the sheet structures except some of the chamosites. These high values are not restricted to the iron-rich chlorites as they are in the other clay minerals. In some instances high R^{3+} occupancy values in the chlorites may indicate that one of the octahedral sheets is dioctahedral rather than trioctahedral.

In most of the analyses reported by Foster, octahedral Al is dominant over octahedral Fe^{3+} although in a few examples the reverse is true. Fe^{3+} is generally high in those samples where Fe^{2+} is nearly as abundant or more abundant than Mg (Fe^{2+}/R^{2+} = 0.45). Kepezhinskas and Sobolev (1965) in a study of chlorites from different rock types and mineral associations, found that in Al-rich chlorites high Fe^{3+} was accompanied by high Fe^{2+} concentrations.

It was thought that the Fe_2O_3 in chlorites was the result of secondary oxidation of FeO; however, Foster found that in many chlorites much of the Fe^{3+} was necessary to maintain a charge balance. The conversion of OH^- to O^{2-} is necessary for the conversion of Fe^{2+} to Fe^{3+}. She found no relation between Fe_2O_3 and O content in excess of 10.0 ions per half cell and concluded that Fe_2O_3 is a normal constituent of many chlorites; this would be particularly true of low-temperature clay chlorites. Hey (1954) divided the chlorites on the basis of more or less than 4% Fe_2O_3. Those with the higher values were considered to be oxidized. Those with the smaller values, which are the most common, were assumed to be unoxidized. (Taylor et al., 1968, showed by use of Mössbauer spectroscopy that some Fe^{3+} can be present in the tetrahedral sheet.)

The chlorite formulas, similar to biotite and trioctahedral clay formulas, indicate that as the Fe^{2+}/Mg ratio increases, the amount of R^{3+} tends to increase. There is a slight reversal of this average trend in the samples with the highest Fe^{2+} content. This association presumably produces the minimum structural strain. When the larger Fe^{2+} substitutes for Mg, the strain is minimized by increased substitution of the smaller trivalent cations. Naturally occurring sheet silicates with a pure Fe^{2+} octahedral sheet apparently do not exist because of the inability of the tetrahedral sheets to adjust to such a large octahedral sheet. Foster's data show that the ratio of Fe^{2+}/R^{2+} ranges from 0.00 to 1.00 indicating a complete range of isomorphous substitution between Fe^{2+} and Mg similar to that in the trioctahedral micas, but in contrast to the relationship in the trioctahedral clays.

A classification of the chlorites was devised by Foster (1962) based on ionic replacement of Al by Si in the tetrahedral sheet and Mg by Fe^{2+} in the octahedral sheet. Fig.17 shows the classification with the location of 150 chlorites. The dividing lines are arbitrary and imply no genetic significance; in fact, they probably have some.

Experimental hydrothermal studies in a portion of the system $MgO-Al_2O_3-SiO_2-H_2O$ showed that a complete sequence of 14 Å chlorites could be synthesized, ranging in composition from amesite $(Mg_4Al_2)(Si_2Al_2)O_{10}(OH)_8$, to penninite $(Mg_{5.5}Al_{0.5})(Si_{3.5}Al_{0.5})O_{10}(OH)_8$ (Nelson and Roy, 1958). A certain minimum amount of R^{3+} substitution is necessary in order to provide sufficient layer charge to bind the various layers. The chlorite structure is not stable with more than one-third of the octahedral positions filled with Al; if more Al is present, a three-phase assemblage is produced, containing a chlorite of the amesite composition.

From the same compositions, but at lower temperatures (below 400–500°C), a 7 Å structure of the kaolinite-type is developed. Nelson and Roy (1958) called these materials septechlorites. It was not established whether the 7 Å phase was metastable or not, but able to persist for long periods as a stable low-temperature polymorphic form of chlorite. They did not synthesize any ferroan chlorites but believed that similar polymorphic relations should exist in the iron chlorites. Turnock and Eugster (1958) synthesized a 7 Å chlorite, daphnite, at a temperature of 400°C with a composition of $(Fe_5Al)(Si_3Al)O_{10}(OH)_8$ but were not able to convert it completely to the 14 Å polymorph.

The structural variations of the chlorites have been discussed by Brindley (1961a), Brown and Bailey (1962, 1963), Shirozu and Bailey (1965), Lister and Bailey (1967), and Eggleston and Bailey (1967). Taking into account the bonding restrictions imposed by the superposition of the 2:1 layers and the hydroxide sheets of ideal hexagonal geometry, Brown and Bailey (1962) showed that four different arrangements of the brucite sheet relative to the initial 2:1 layer were possible. For each of the four chlorite layer types, the hexagonal rings in the repeating 2:1 layer may be superimposed in six different orientations. Six ideal systems with semi-random stacking are possible, two (Ib and IIa) based on an orthorhombic-shaped cell and four (Ia, Ib, IIa, IIb) based on a monoclinic-shaped cell. Detailed structure analyses by Steinfink (1958a,b), Brown and Bailey (1963), and Shirozu and Bailey (1965) indicate that the 2:1 network is distorted by tetrahedral rotation. In part, this rotation allows the tetrahedral sheet to adjust to the size of the larger (b-axis) octahedral part of the 2:1 layer; in addition, the direction of rotation is such that a shorter and more favorable hydrogen bond system between the talc layer and the brucite sheet is formed.

In three of the four structures that were determined in detail the 2:1 octahedral sheet and the hydroxide octahedral sheet show a difference in cation population. In three of these specimens all or most of the octahedral charge (R^{3+}) is located in the hydroxide sheet. In the other the positive charge is split between the two octahedral sheets. In two of these chlorites (Steinfink, 1958a; Brown and Bailey, 1963) the Al in the tetrahedral sheet appears to have an orderly rather than a random distribution.

Brown and Bailey (1962) examined 300 chlorites from different localities and found that approximately 80% had the IIb structure but found examples of the orthorhombic Ib, monoclinic Ib and Ia structural type. The relative abundance of the polytypes was related to structural stability. Composition influences to some extent the stability of the chlorites through its effect on the cation charge and amount of distortion of the hexagonal network caused by size adjustments. Increasing tetrahedral Al substitution is accompanied by an increase in octahedral Fe to maintain a reasonable degree of fit between the two types of layers.

There is considerable overlap in the compositions of the various structural polytypes. The Ia and Ib chlorites have relatively distinct compositions, both being low in tetrahedral Al. Average compositions as calculated by Brown and Bailey (1962) are:

Polytype	Number of specimens	Average tetrahedral Al	Average Fe/Fe+Mg
IIb, $\beta = 97°$	111	1.31	0.39
Ib, $\beta = 90°$	36	1.32	0.59
Ib, $\beta = 97°$	13	0.87	0.52
Ia, $\beta = 97°$	4	1.03	0.38

The IIb chlorite is the stable polytype in normal chlorite-grade metamorphism and in medium and high-temperature ore deposits. Brown and Bailey suggest that when sufficient energy is available, the most stable polytype (IIb) will form. They found that the orthohexagonal and monoclinic Ib types were the ones most likely to be considered diagenetic chlorites. These are the chlorites with the lowest amount of tetrahedral Al and these polytypes are most apt to be stable with a small amount of tetrahedral Al. The Ib types readily convert to the more stable IIb types during metamorphism. In some specimens increasing Fe and tetrahedral Al content seems to correlate with increasing temperature of formation. A similar vague relation exists for the clay minerals.

In a study of chlorite in sedimentary rocks, Hayes (1970) concluded that type-I chlorite most likely represents authigenic chlorite (because of its relative instability); the IIb stable polytype, in most cases, would indicate that the chlorite is detrital and reflects formation by igneous or metamorphic processes. Hayes points out that a few of his samples of IIb chlorite appear to be authigenic, probably formed in a higher-temperature environment caused by deep-burial or hydrothermal activity.

Trioctahedral clay chlorite

Chlorites are a common constituent of low-grade metamorphic rocks. They are less common in igneous rocks where they occur as hydrothermal alteration products of ferromagnesium minerals. Chlorite occurs in nearly all types of sedimentary rocks and though it is seldom the dominant sheet silicate present, it probably occurs in something like 75–90% of sedimentary rocks. Most of the chlorite in shales and sands is detrital in origin, particularly in the graywacke facies, where it is derived directly from low-grade metamorphic rocks and usually has suffered little weathering. Chlorite is quite easily weathered under moderate acid leaching conditions and it is a little surprising that so much of it survives in the sedimentary record. Initially the hydroxide sheet is stripped from chlorite and a vermiculite or mixed-layer chlorite-vermiculite is formed. Al and Fe^{3+} hydroxide are commonly precipitated between the expanded layers which tends to retard the weathering process. This is a relatively stable product and apparently much of it persists long enough to be transported to the sea; there, additional hydroxides may be obtained and the clay reverts to chlorite. Also, it is likely that acid weathering was of less importance in the geologic past than at present.

It is difficult to obtain sedimentary chlorites of sufficient purity for reliable chemical analyses; however, reasonably accurate estimates of chemical composition can often be made from X-ray data. If the difference in scattering power and size of the various atoms in the chlorite species is sufficiently large, the positions and intensities of X-ray reflections are measurably affected. The b and c parameters are the most useful.

The 060 reflections of trioctahedral chlorite range from 1.53 to 1.55 Å and vary linearly with compositional variations in the octahedral sheet. Two formulas for the b parameter are:

$b = 9.210 + 0.037 (Fe^{2+}, Mn)$ (Shirozu, 1958)
$b = 9.23 + 0.03 Fe^{2+} \pm 0.0285$ (Radoslovich, 1962)

As might be expected, Fe^{2+} is considerably more important than the smaller Fe^{3+} and Al ions.

The basal spacing depends on the dimensions of the 2:1 layer and the hydroxide sheet and the forces holding them together. This bonding force is a function of the amount of charge in the tetrahedral and octahedral sheets (particularly the hydroxide sheet) and is therefore proportional to the amount of Al substitution in the tetrahedral sheet (Brindley,1961a). Brindley has given the following linear relation: $d(001) = 14.55 - 0.29\ x$ (x represents the amount of tetrahedral Al). He suggests a correction factor of $-0.05\ Fe^{3+}$ may be necessary for leptochlorites (greater than 4% Fe_2O_3 or 0.4 Fe^{3+}). Brindley has also described a method (after Brindley and Gillery,1956) of calculating the Fe content and the difference in composition of the two octahedral sheets using the intensities of the first five peaks. The common presence of vermiculite layers and faults in clay-sized chlorites make the application of this technique difficult. In this same article Brindley and Gillery show how variations in basal intensities can be used to identify dioctahedral chlorites and chlorites with dioctahedral mica-like layers and trioctahedral brucite sheets. The most diagnostic feature of these chlorites is the presence of a relatively strong 003 reflection.

The 001 reflections (14.25–14.05 Å) of most chlorites in shales indicate that they have from 1.0 to 1.9 tetrahedral Al per four tetrahedral positions. The tetrahedral Al in the majority of these chlorites falls within the narrower range of 1.0–1.6 (14.25–14.10 Å). An average value would be close to 1.4. (Most of the macrochlorites lie within a similar range.) Available 060 values are limited; however, the relative intensities of the 001 reflections indicate that most shale chlorites have from 25 to 67% (1.5–4.0) of the octahedral positions filled with iron and that, in many samples, there is more Fe in the hydroxide sheet than in the 2:1 layer. The Fe values are probably high. The presence of kaolinite and faults in the chlorite would tend to give anonymously high values for Fe. This is believed to be the case for many samples. The presence of vermiculite layers, which would give low tetrahedral Al values and relatively low Fe values, is another source of error. Chemical analyses of five chlorite-rich (50–90%) samples concentrated from limestones, shales and sandstones gave iron oxide values ranging from 7 to 17%. Assuming all of the iron is in the chlorites, they would contain between 10 and 20% iron oxides of which the large proportion is in the ferrous form. This amount of iron would fill approximately 0.8–1.7 octahedral positions or considerably less than was suggested by the relative intensities of the odd and even order reflections.

Some of the high Fe values may be real. During weathering under neutral to acid conditions, the Mg-rich brucite sheet tends to be stripped out and removed from the immediate environment. If new material is precipitated between the talc layers, it is more apt to be Fe and Al than Mg. As chlorites go through the sedimentary cycle, perhaps several times, their average Fe and Al content will tend to increase.

Many high-iron chlorites associated with sedimentary iron formations apparently have formed by direct precipitation. These have compositions similar to chamosite (see Table LXXII) and occur with both a 7 Å and 14 Å structure (Brindley, 1961a).

Schoen (1964) calculated from X-ray data that the secondary chlorite in the Clinton ironstones had the following composition: $(MgAl)_{2.5}(Fe)_{3.5}(Si_{2.65}Al_{1.35})O_{10}(OH)_8$, with the 2:1 layer having more Fe than the hydroxide sheet. Co-existing, primary 7 Å chamosite from which much of the chlorite formed has a similar composition.

Secondary, post-depositional chlorite is a common component in marine sands, where it commonly forms from the alteration of volcanic material or montmorillonite. X-ray analyses suggest that many of these chlorites are relatively iron-rich. A partial analysis (Weaver and Beck, 1972) of a clay chlorite from a Pennsylvanian sandstone from Oklahoma tended to confirm the idea that these clays are rich in iron: 28% SiO_2, 15.0% Al_2O_3, 23.9% Fe as Fe_2O_3, 5.3% MgO, 2.6% CaO, 1.2% K_2O.

Trioctahedral clay chlorite is an abundant constituent of soils formed by the weathering of basic volcanic pumice and tuffs in North Wales (Ball, 1966). The adjusted chemical analysis (29.35% SiO_2, 16.82% Al_2O_3, 4.42% Fe_2O_3, 15.08% FeO, 0.25% MnO, 21.54% MgO, 12.00% H_2O^+, 0.54% H_2O^-) produces the following structural formula:

$$\overset{+0.91}{(Al_{0.95}Fe^{3+}_{0.34}Fe^{2+}_{1.27}Mn_{0.02}Mg_{3.25})_{5.81}} \overset{-1.04}{(Si_{2.96}Al_{1.04})O_{9.94}(OH)_{8.06}}$$

The formula calculated from X-ray data is:

$$(Al_{1.21}(Fe, Mn)_{1.50}Mg_{3.29})_{6.00}(Si_{2.79}Al_{1.21})O_{10}(OH)_8$$

The X-ray method affords a reasonable estimate of the structural formula based on chemical data. Using X-ray data, Ball calculated the structural formulas for twenty-six weathered soil and vein clay chlorites from North Wales. Tetrahedral Al ranged from 1.0 to 1.7 which is similar to the values for chlorites in shales. Octahedral Fe ranged from 0.8 to 2.4, with all but two values being less than 1.6; these values are much lower than those calculated (X-ray) for shale samples but almost identical to the shale values based on chemical determination of the Fe content.

Not only can trioctahedral chlorites form in soils but there appears to be relatively little difference in composition between them and chlorites in sediments. This indicates that the trioctahedral chlorites in sediments can not automatically be assumed to be detrital.

Mg-rich chlorites do not seem to form readily in low-temperature marine environments. Mixed-layer chlorite-vermiculites form fairly easily in magnesium-rich environments but complete development of the brucite sheets must be considerably more difficult. Mixed-layer chlorite-vermiculite is the predominant clay in the Lower Ordovidian limestones and dolomites of southern United States, usually over a thousand feet thick and extending over 500,000 square miles, yet very little chlorite is present (Weaver, 1961a). These mixed-layer clays and others like them appear to have formed on extensive tidal flats where the clays were exposed to alternating evaporitic and normal marine conditions. The original source material may have been volcanic material and/or montmorillonite. Carstea et al. (1970) formed relatively stable hydroxy-Mg interlayers in montmorillonite after 10 days equilibration at pH 10.4. Some interlayers also formed between pH 6.8 and pH 9.8. Less stable interlayers were formed in vermiculite at pH 10.7 and no detectible interlayers formed at pH 9.7. In both the montmorillonite and vermiculite the amount of interlayers decreased with time, the vermiculite losing the interlayers at a faster rate.

Dioctahedral clay chlorite

Most of the chlorite-like material formed in soils is dioctahedral rather than trioctahedral. In the process of weathering, illite and muscovite are stripped of their potassium and water enters between the layers. In these minerals and in montmorillonites and vermiculites, hydroxides are precipitated in the interlayer positions to form a chlorite-like mineral (Rich and Obenshain, 1955; Klages and White, 1957; Brydon et al., 1961; Jackson, 1963; Quigley and Martin, 1963; Rich, 1968). $Al(OH)_3$ and $Fe(OH)_3$ are likely to be precipitated in an acid to mildly basic environments and $Mg(OH)_2$ in a basic environment. The gibbsite sheets in the soil chlorites are seldom complete and the material resembles a mixed-layer chlorite-vermiculite. The gibbsite may occur between some layers and not between others or may occur as islands separated by water molecules.

Lazarenko (1940) described a hydrothermal aluminum-silicate mineral from the Donetz Basin which he called donbassite. This material is a variety of dioctahedral chlorite and the Nomenclature Committee (Bailey et al., 1971) considers that it has priority.

Caillère and Hénin (1949) and Slaughter and Milne (1958), using a concentrated solution of Al, have caused interlayer Al hydroxide (as well as Mg and other hydroxides) sheets to deposit between montmorillonite layers. Hsü and Bates (1964) were able to deposit hydroxy-aluminum polymers between layers of vermiculite as long as the NaOH/Al ratio was less than 3. At higher values, crystalline $Al(OH)_3$ was formed outside the clay structure. Barnhisel and Rich (1963) found that Al interlayers in montmorillonite were not stable and on aging would crystallize out as the mineral gibbsite. Using solutions with an OH/Al molar ratio ranging from 1.50 to 3.00 they showed that stability increased as molar ratio decreased. As the composition approaches $Al(OH)_3$, the charge approaches neutrality and attraction for the 2:1 layer decreases and stability decreases. Even though gibbsite normally forms under neutral to alkaline conditions, Barnhisel and Rich were able to form gibbsite in the presence of montmorillonite at a pH of less than 4.3. Montmorillonite is presumed to act as an anion, repelling other large anions and allowing the interlayer Al to hydrolyze to $Al(OH)_3$. Polymerization then occurs. The interlayer material is nearly neutral and in the presence of an exchange cation gibbsite will be "split off" from the montmorillonite layer. It is likely that the stability of the Al interlayers (and Mg, Fe, etc.) and the reaction of the various molar ratios are to a large extent controlled by the charge and a, b dimensions of the 2:1 template.

Carstea (1968) found that the formation of hydroxy-Al in the interlayer position of vermiculite and montmorillonite was temperature-dependent, the amount increasing with increasing temperature. The formation of interlayer hydroxy-Fe was not temperature-dependent for montmorillonite and only slightly so for vermiculite.

Although partially organized dioctahedral chlorites form readily in soils, there are relatively few reported in sedimentary rocks. (More dioctahedral chlorites probably exist than have been recognized.) Swindale and Fan in 1967 reported the alteration of gibbsite deposited in Waimea Bay off the coast of Kauai, Hawaii, to chlorite but no data were obtained on its composition. Dioctahedral chloritic clays have been reported forming in recent marine sediments; however, the identification is indirect and the interlayer material is relatively sparse (Grim and Johns, 1954).

Weaver (1959) noted that the chlorite, which is a common constituent of the Ordovician K-bentonite beds of the eastern United States, has a dioctahedral 2:1 layer and a trioctahedral hydroxide sheet. A partial chemical analysis indicated the chlorite contained less than 2% Fe_2O_3. Both layers were probably formed in place from the alteration of volcanic ash in a marine environment. Only one other chlorite of this type had been detected in X-ray patterns of approximately 75,000 samples of sedimentary rocks. The other sample was from a Paleozoic argillaceous limestone at a depth of 24,400 ft. in Oklahoma. Chlorites of this type might well go undetected when chlorite is only a minor component.

Eggleston and Bailey (1967) published a study on dioctahedral chlorite and gave five examples of chlorites having a pyrophyllite-like layer and a brucite-like sheet (designated di/trioctahedral by the authors with the trioctahedral sheet including all species of chlorite with 5 to 6 octahedral cations per formula unit and dioctahedral 4 to 5 octahedral cations per formula unit). Identification of di/trioctahedral chlorites is indirectly accomplished. Eggleston and Bailey stated that identification "depends on the intermediate value of $d(060)$, on chemical analysis of impure material, and on the ideal compositions of the recrystallization products of static heating". The composition of one such chlorite for which they refined the structure is:

$$[Al_{2.0}(Si_{3.3}Al_{0.7})O_{10}(OH)_2]^{-0.7}[(Mg_{2.3}Al_{0.7})(OH)_6]^{+0.7}$$

They explained the fit of the dioctahedral and trioctahedral sheets by the thinning of the dioctahedral sheet (2.05 Å) and the thickening of the trioctahedral sheet (2.15 Å), giving a mean lateral octahedral edge of 3.02 in both sheets. To compensate for the thinning of the dioctahedral sheet the tetrahedral sheet thickened slightly. The tetrahedral rotation is in the same direction as for the other chlorites, but the angle is somewhat less than expected.

TABLE XLII

Chemical analyses and structural formulas of some dioctahedral chlorites

	1	2	3	4	5	6	7
SiO_2	39.85	29.5	38.0	36.7	39.05	35.63	39.01
Al_2O_3	33.52	52.0	47.4	49.5	42.49	34.87	32.15
Fe_2O_3	4.56	0.5	–	–	–	5.01	0.90
FeO	–	0.3	–	–	0.68	0.43	0.10
MgO	4.58	–	0.8	0.0	2.18	8.63	10.14
CaO	0.15	1.6	–	–	–	–	–
Na_2O	–	–	–	–	–	0.24	0.10
K_2O	1.61	–	–	–	–	0.46	1.52
TiO_2	1.03	1.2	–	–	0.25	none	0.47
H_2O	13.18	13.9	13.8	13.8	14.17	14.15	14.15
Total	98.48	99.0[a]	100.0	100.0	100.00[b]	100.60	99.50[c]

1. Brydon et al. (1961): clay fraction of the AB horizon of the Alberni soil series in British Columbia; the formula is approximate as it was necessary to make a number of assumptions in the calculation.
2. Caillère et al. (1962): in bauxite deposits of Pyrenées orientales.
3,4. Müller (1963): hydrothermal alteration of acid tuff from Kesselberg, Treiberg, Germany.
5. Heckroodt and Roering (1965): mixed-layer chlorite-swelling chlorite; hydrothermal alteration of beryl, Karibib, Southwest Africa; analysis recalculated to 100%; analysts P.J. Fourie, C.E.G. Schutte and A.F. Hogg.

A number of Al chlorites in which both octahedral sheets are dioctahedral have recently been described. Dioctahedral Al chlorites have been reported in bauxite deposits (Bardossy, 1959; Caillère, 1962). These chlorites appear to have been formed by the "precipitation-fixation" of Al hydroxide in the interlayer position of stripped illite or montmorillonite. A similar type of chlorite, along with dioctahedral chlorite-vermiculite, occurs in the arkosic sands and shales of the Pennsylvanian Minturn Formation of Colorado (Raup, 1966). Bailey and Tyler (1960) have described the occurrence of dioctahedral chlorite and mixed-layer chlorite-montmorillonite in the Lake Superior iron ores. Hydrothermal occurrences have been described by Sudo and Sato (1966).

Chemical analyses of dioctahedral chlorites have been given by Brydon et al. (1961), Caillère (1962), Müller (1963), Heckroodt and Roering (1965), and Sudo and Sato (1966) — Table XLII. Two of these have a sedimentary origin and five a hydro-

	1	2	3	4	5	6	7
Tetrahedral							
Si	2.8	2.70	3.3	3.2	3.40	3.26	3.24
Al	1.20	1.30	0.7	0.8	0.60	0.74	0.78
Octahedral							
Al	3.22	4.3	4.17	4.27	3.75	3.01	3.04
Fe^{3+}	0.38	0.03	–	–	–	0.34	0.04
Fe^{2+}	–	0.02	–	–	0.05	–	–
Mg	0.76	–	0.10	–	0.28	1.18	1.59
Σ	4.36	4.35	4.27	4.27	4.27[d]	4.53	4.67
Interlayer							
K	0.23	–	–	–	–	–	0.16
Na	–	–	–	–	–	–	0.02
Ca	0.02	–	–	–	–	0.11	0.06
X	0.24	–	–	–	–	–	–

6. Sudo and Sato (1966): associated with hydrothermal pyrite ore body, Aomori, Japan.
7. Sudo and Sato (1966): hydrothermally altered argillaceous material around gypsum deposit, Akita, Japan.

[a]includes 1.0% P_2O_5; [b]includes 1.18% Li_2O; [c]includes 0.42% S; [d]includes 0.02% Ti and 0.41% Li; based on determined H_2O^+, 09.78, $OH_{8.22}$

thermal origin. The 060 value is slightly less than 1.50 Å and the 001 value of the non-swelling specimens is approximately 14.2 Å.

In order to maintain electrical neutrality the octahedral sheets must contain more than 4 Al per $O_{10}(OH)_8$. The excess octahedral charge should equal the amount of negative tetrahedral charge if no exchange cations are present. Reported tetrahedral charges range from 0.60 to 1.30. These charges are fully satisfied by excess octahedral occupation in the hydrothermal samples, but not in the sedimentary samples. The sedimentary specimens have a much higher tetrahedral Al content (1.20 and 1.30) than the hydrothermal clay and if the analyses are reasonably accurate, it would suggest that the basic 2:1 layers were not originally stripped illite or montmorillonite. It may be that the starting material was stripped biotite or chlorite from which much of the octahedral Mg and Fe was removed and, in part, replaced by Al. Such replacement would be favored in an acid-soil environment (see Chapter 7: Vermiculite).

The hydrothermal dioctahedral chlorites have considerably less tetrahedral substitution than those formed in sediments. The former would appear to be a stable phase and the latter a metastable phase. The tetrahedral composition of the hydrothermal specimens is similar to that for the other dioctahedral clays and represents a reasonable fit between the tetrahedral and octahedral sheets.

Biotite is abundant in crystalline rocks although relatively little of it is found in sediments. Some of it is altered to form trioctahedral chlorites and mixed-layer chlorite-vermiculites. It is likely that much of it is further altered to form dioctahedral chlorites and mixed-layered chlorite-montmorillonites (or vermiculite). The tetrahedral sheet of high-temperature biotite (and chlorite) has a relatively large amount of Al and is larger (in the a,b direction) than the tetrahedral sheets of other common 2:1 minerals. Considerable strain would be induced in an attempt to fit dioctahedral octahedral sheets to such large tetrahedral sheets. In part, this misfit is compensated for by more than two octahedral positions being filled; however, it is unlikely that this is sufficient to reduce the strain enough to allow these clays to develop any degree of size or stability. The unstable and discontinuous interlayer gibbsite sheet could be partially due to its inability to adjust to the size of the large tetrahedral sheet. Thus, the stability and crystallinity of secondary chlorites, formed by the deposition of material in the interlayer spaces, would be a function of the composition and size of the original 2:1 layer and the type of cations available to fill the interlayer positions. The latter, to some extent, would be a function of the pH conditions.

Some chlorite appears to form in muds and shales during deep burial. Much of this chlorite may be dioctahedral formed by Al released from kaolinite and feldspar (Weaver and Beck, 1971a).

Chapter 7

VERMICULITE

Macroscopic vermiculite

Like chlorite most of our understanding of vermiculite is based on a study of macroscopic crystals. Vermiculite occurs as a clay-sized material in sediments although much of it occurs as a mixed-layer clay. Vermiculite is somewhat of a waste-basket term. Basically, the name is given to the 2:1 expandable minerals which generally have a layer charge larger than that of smectite (~0.6 per $O_{10}OH_2$). Vermiculite can be either trioctahedral or dioctahedral. Most macroscopic vermiculites are trioctahedral, whereas both types are common as clay-size minerals. Most vermiculites form by the degradation of pre-existing sheet silicates and much of their character is determined by that of the starting mineral. Thus, it is not surprising that a restricted definition of vermiculite has not been agreed upon.

The expanded 2:1 minerals have a continuous gradation of cation exchange capacities ranging from approximately 25 to 250 mequiv./100 g. Weiss et al. (1955) and Hofmann et al. (1956) suggest a division be made at 115 mequiv./100 g. This is equivalent to a layer charge of 0.55 per $O_{10}(OH)_2$ unit of structure; Walker (1958) has suggested the boundary value should be placed slightly higher. An average layer charge of 0.41 was obtained for the 101 montmorillonite analyses used in this study. This would suggest that a boundary value of 0.55 is low.

Why is there a need for a boundary? As the particle size, total amount of layer charge, and relative amount of tetrahedral charge vary, the expansion, contraction, and adsorption properties of expandable 2:1 clays vary. In general, as these three properties increase, and they frequently (but not always) increase together, the layers are better able to contract to 10Å when potassium is placed in the interlayer positions; they will imbibe one layer of ethylene glycol or glycerol rather than two. The interlayer water is better organized and fewer layers can normally be accommodated in the interlayer position (Walker, 1961). These processes generally occur when the total layer charge is larger than 0.6 to 0.7. Larger flake size and a higher proportion of tetrahedral charge should lower this value slightly. Macroscopic vermiculites generally have all the required properties and identification is relatively straight-forward. Most classification problems occur with clay-sized vermiculites. Much of the problem is intimately related to the origin of the expandable clays.

Most macroscopic vermiculites have formed by the removal of potassium from biotite. Biotite can easily be altered to vermiculite in the laboratory by exposing it to a dilute Mg solution; however, Bassett (1959) has shown that if as little as 0.04 M potassium is present, biotite will not alter to vermiculite. Biotites have a theoretical exchange capacity of 250–260 mequiv./100 g. Oxidation of octahedral iron reduces this somewhat but the resulting charge is generally significantly larger than that of smectite and usually enough to cause layer contraction when potassium is accessible. Muscovites that are leached of potassium would tend to have an even higher charge as these contain less ferrous iron. Expanded layers are also formed by alteration of such minerals as chlorite, serpentine and hornblende, with chlorite being by far the most important. The layer charge produced by the alteration of these minerals is considerably less than that produced by the alteration of the micas. The vermiculites produced from these latter minerals in general do not contract when exposed to potassium and generally have a layer charge less than 0.7. The layer charge range of these minerals overlaps that of the fine grained smectites; however, the vermiculites are usually coarser and some tend to have a higher proportion of tetrahedral charge than some finer grained expanded clays (montmorillonite). It is at this stage that assigning names becomes difficult and somewhat meaningless. In the literature, the use of the names *montmorillonite* and *vermiculite*, when referring to the expanded clays in the range of high-charged smectite and low-charged vermiculite, particularly when they occur interlayered with chlorite, is quite arbitrary. Weaver (1958) has suggested that, for the time being, an empirical subdivision be made on the basis of whether or not an expanded clay will contract to 10 Å when placed in a potassium solution. Walker (1957) has suggested the distinction be made on the basis of whether the layers adsorb one or two layers of glycerol. The former technique places the boundary at a slightly higher charge value.

Foster (1963) calculated the structural formulas for 25 macroscopic vermiculites (Table XLIII). These vermiculites have octahedral compositions similar to phlogopites, Mg-biotites, Mg-chlorite, and saponite, but due to the oxidation of much of the iron, tend to have a higher proportion of trivalent octahedral cations. The tetrahedral sheets have the relatively high-charge characteristic of biotites and chlorites. Octahedral R^{3+} occupies up to 40–45% of the filled octahedral positions; this is a higher value than is found for any other trioctahedral sheet structure clay except for the high iron 1:1 clays. In these clays most of the high R^{3+} values are probably due to the oxidation of iron during weathering.

Taylor et al. (1968) studied iron in two vermiculites by Mössbauer spectroscopy. They found all the iron in both samples to be in the Fe^{3+} oxidation state. The ferric iron showed no preference for a particular octahedral site; in fact, some Fe^{3+} was present in tetrahedral sites.

Most of the vermiculites listed by Foster apparently were formed by the leaching of K from biotite. Biotite has a negative layer charge near 1.00 per $O_{10}(OH)_2$ units. Foster's vermiculites have charges from 1.08 to 0.38 with only five of 25 values

TABLE XLIII

Partial structural formulas for some vermiculites (selected from Foster, 1963)

	1	2	3	4	5	6	7	8
Octahedral								
Al	0.00	0.15	0.00	0.00	0.16	0.25	0.35	0.36
Fe^{3+}	0.17	0.01	0.14	0.39	0.28	0.09	0.44	0.87
Fe^{2+}	0.06	–	0.02	0.01	0.04	0.40	0.10	0.19
Mg	2.77	2.83	2.84	2.60	2.45	2.26	2.11	1.40
	3.00	2.99	3.00	3.00	2.93	3.00	3.00	2.82
Layer charge								
octahedral	+0.17	+0.14	+0.14	+0.39	+0.30	+0.34	+0.80	+0.87
tetrahedral	−1.13	−1.14	−0.94	−1.14	−1.07	−1.34	−1.22	−1.39
total	−0.94	−1.00	−0.80	−0.75	−0.77	−1.00	−0.42	−0.52

being less than 0.57. The overlap of smectite layer charge values is relatively small. The reduced layer charge is due to the oxidation of ferrous iron. This oxidation weakens the potassium bonds and allows the potassium to be replaced, usually by the divalent Mg ion, and the layer then expands. Because a minor amount of potassium in solution will prevent the replacement of K by Mg (Basset, 1963) efficient leaching conditions must be necessary for the formation of vermiculite from biotite. D'yakonov and L'vova (1967) interpreted their data, obtained from treating biotites containing Fe^{3+} oxidized during weathering with $MgCl_2$ solutions, as showing that the oxidation of Fe^{2+} to Fe^{3+} causes potassium to be more tightly bound between the layers, thus preventing vermiculite from forming. The potassium would be held because of the loss of hydroxyl protons and/or the substitution of F^- for OH^- when Fe^{2+} is oxidized to Fe^{3+}.

High-iron biotites are not likely to alter to vermiculite. When the ferrous iron content is sufficiently high, oxidation will result in a positive layer charge and the extension of the octahedral sheet to such an extent that the basic layer would be unstable and break up quite rapidly.

Foster (1963) established empirically that the cation exchange capacity of vermiculites could be calculated by multiplying the positive charges carried by the interlayer cations by 200. The calculated range of cation exchange capacities of the macroscopic vermiculite she studied ranged from approximately 80 to 200 mequiv/100 g. Nearly half the samples have a C.E.C. less than 140 and more than a third have values less than 120. Because it was not known what proportions of the Mg to assign to the interlayer position and what proportion to the octahedral sheet, these calculated values can only be considered minimum values. Macroscopic vermiculites most commonly contain Mg in the interlayer position. Minor amounts of Ca, Na, and K are present in many samples with Ca being the most common.

Curtis et al. (1969) described a naturally occurring vermiculite with Na as the major interlayer cation. They reported that the Na-vermiculite was similar to synthetically produced Na-vermiculite by the exchange of Na for K in phlogopite. The noteworthy difference between the two was the lower layer charge in the naturally occurring vermiculite, the suggested reason being that the larger amount of ferric iron represents oxidation. This would lower the layer charge to a value close to that of the macroscopic Mg-vermiculites.

The large size of the vermiculite flakes allows the nature of the interlayer material to be defined more specifically than for the clay-sized minerals. Mathieson and Walker (1954) and Mathieson (1958) have shown that the interlayer water consists of two discrete sheets. The water molecule sites within each sheet are arranged in a near hexagonal array with each site being equivalently related to a single oxygen in the slightly distorted silicate layer surface. Only about two-thirds of the available water molecule sites and one-ninth of the exchangeable cation sites were occupied in the specimen Mathieson and Walker examined. This is equivalent to approximately one Mg per one and one-half unit cells versus six Mg per unit cell for a brucite sheet. The Mg ions lie between the water sheets and are octahedrally coordinated with the water molecules. They point out that their model assumes static conditions. There is actually a constant migration of the interlayer water molecules and cations at normal temperatures.

Clay vermiculite

Vermiculite and vermiculite layers interstratified with mica and chlorite layers are quite common in soils where weathering is not overly aggressive. (A few references are: Walker, 1949; Brown, 1953; Van der Marel, 1954; Hathaway, 1955; Droste, 1956; Rich, 1958; Weaver, 1958; Gjems, 1963; Millot and Camez, 1963; Barshad and Kishk, 1969.) Most of these clays are formed by the removal of K from the biotite, muscovite and illite and the brucite sheet from chlorite. This is accompanied by the oxidation of much of the iron in the 2:1 layer. Walker (1949) has described a trioctahedral soil vermiculite from Scotland formed from biotite; however, most of the described samples are dioctahedral. Biotite and chlorite with a relatively high iron content weather more easily than the related iron-poor dioctahedral 2:1 clays and under similar weathering conditions are more apt to alter to a 1:1 clay or possibly assume a dioctahedral structure.

Some of expanding clays in soils have the attributes of vermiculite, some of smectite, and some have features of both. The variation in properties is largely related to the layer charge. The charge is dependent on the original charge on the 2:1 layer of the parent mineral and the amount of ferrous iron in the octahedral sheet. The oxidation of this iron reduces the layer charge. The layer charge of the dioctahedral starting materials ranges from 0.6 to 1.0 per $O_{10}(OH)_2$. (Ferrous iron values can be as high as 0.65 per two octahedral positions.) Thus, it can be expected that clay vermiculites would vary widely in composition and charge and not necessarily bear much relation to the macroscopic vermiculites. Chemical analyses of pure clay vermiculite appear to be non-existent; however, Barshad and Kishk (1969) made a detailed mineral and chemical study of 11 soils and calculated structural formula for the vermiculitic component (Table XLIV).

They found that the soil vermiculites could be placed into two groups. The first group had at least one Al per four tetrahedral sites and Al in the octahedral sheet (No.1–5). The charge deficiency in the tetrahedral sheet was partially compensated for by a positive charge in the octahedral sheet (two cations per three sites). These samples are similar to the coarser-grained vermiculites in composition. The second

TABLE XLIV

Structural formulas for vermiculite clays (After Barshad and Kishk, 1969)

	1	2	3	4	5	6	7	8	9	10	11
Octahedral											
Al	1.44	1.47	1.24	0.93	1.12	0.20	0.43	0.33	–	0.88	–
Fe^{3+}	0.16	0.21	0.20	0.44	0.29	0.55	0.36	0.49	0.47	0.42	0.90
Ti	0.14	0.13	0.18	0.24	0.16	0.15	0.05	–	–	–	–
Mg	0.27	0.25	0.35	0.46	0.75	1.11	1.52	1.26	1.91	0.66	1.22
Mn	0.05	–	0.02	0.07	0.03	–	–	0.08	0.02	0.08	0.03
H	0.30	0.34	0.25	0.26	0.19	0.27	0.21	0.19	0.08	0.04	0.04
Σ	2.36	2.40	2.24	2.40	2.65	2.28	2.57	2.22	2.44	2.06	2.17
Tetrahedral											
Al	1.10	1.13	0.76	1.11	1.31	0.18	0.60	–	–	–	–
Si	2.90	2.87	3.24	2.89	2.69	3.82	3.40	4.0	4.0	4.0	4.0
Interlayer											
Na	0.43	0.55	0.52	0.57	0.44	0.60	0.42	0.68	0.65	0.58	0.76
K	0.37	0.18	0.21	0.15	0.25	0.24	0.36	–	–	–	–
Layer charge											
octahedral	+0.30	+0.40	+0.03	+0.39	+0.62	−0.66	−0.18	−0.68	−0.65	−0.58	−0.76
tetrahedral	−1.10	−1.13	−0.76	−1.11	−1.31	−0.18	−0.60	0.0	0.0	0.0	0.0
total	0.80	0.73	0.73	0.72	0.69	0.84	0.78	0.68	0.65	0.58	0.76
C.E.C. (mequiv./100 g)	212	194	181	198	181	207	199	168	174	151	196

LEGEND TABLE XLIV

Structural formulas for vermiculite clays (After Barshad and Kishk, 1969)

Sample No.	Soil series and Ref. No.	Great soil group	Location	Parent material	pH of soil	Mode of formation of vermiculite clay
1	Taiwan 11A	lateritic red earth	eastern Taiwan, China	fine grain schist	5.7	alteration of mica
2	Josephine 62-12-24-4	gray brown podzolic	Humboldt Co., Calif., U.S.A.	sandstone and shale	5.4	alteration of mica
3	Melbourne 61-12-31-5	brown podzolic	Humboldt Co., Calif., U.S.A.	sandstone and shale	5.2	alteration of mica
4	Shaver 59-10-18-1	podzolic	Fresno Sierra, Calif., U.S.A.	quartz, diorite	6.6	alteration of mica
5	Masterson 58-52-22-3	podzolic	Tehama Co., Calif., U.S.A.	mica, schist	5.3	alteration of mica
6	Neuns 58-52-14-3	podzolic	Tehama Co., Calif., U.S.A.	greenstone	6.4	alteration and synthe
7	Yorkville 54-12-1-3	prairie	Humboldt Co., Calif., U.S.A.	chloritic, greenstone	6.0	alteration and synthe
8	Aiken 13B	red podzolic	Lytonville Calif., U.S.A.	greenstone	6.0	alteration and synthe
9	Sweeney 18-24	prairie	San Mateo Co., Calif., U.S.A.	basalt	6.4	synthesis
10	Auburne 0-8	non calcic brown	Merced Co., Calif., U.S.A.	greenstone	6.0	synthesis
11	Boomer 58-52-26-7	brown forest	Tehama Co., Calif., U.S.A.	basalt	5.7	synthesis

groups had no tetrahedral Al, the charge deficiency arising from the octahedral sheet in which Mg and Fe^{3+} were the dominant cations (No.8–11). Samples 6 and 7 are mixtures of the two forms of vermiculite clay. Both groups were more dioctahedral than trioctahedral and had similar interlayer charges (0.60–0.80). The vermiculite with Al in the tetrahedral sheet is associated with mica and formed from the alteration of mica derived from acid igneous rocks and their sedimentary derivatives. The samples in the group with no Al in the tetrahedral sheet and dominant Mg and Fe^{3+} in the octahedral sheet are associated with montmorillonite and probably formed by synthesis from primary oxides of Si, Al, Fe, and Mg with a basic igneous parent material. Barshad and Kishk suggested that this group of vermiculites forms a continuous series with montmorillonite distinguishable from the montmorillonite only by a determination of C.E.C.

Charge on clay vermiculites usually varies as a function of flake size. With decreasing size and increasing total surface area and edge area, weathering is more effective and the layer charge is reduced. In some instances the coarser fraction resembles a typical vermiculite and the finer fraction a typical montmorillonite. This has been substantiated in work done by Kerns and Mankin (1967) who separated a weathered vermiculite into eight size classes ranging from 0.05 micron to 32 microns. Calculated structural formulas indicate a systematic decrease in tetrahedral Al and octahedral Mg accompanied with an increase in tetrahedral Si and octahedral Fe and Al:

$$16-32\mu \quad K_{0.01}Ca_{0.05}Mg_{0.32}(Al_{0.04}Fe^{3+}_{0.38}Mg_{2.31})_{2.77}(Si_{3.27}Al_{0.76})O_{10}(OH)_2$$

$$0.10-0.25\mu \quad K_{0.03}Ca_{0.04}Mg_{0.11}(Al_{0.44}Fe^{3+}_{0.85}Mg_{1.02})_{2.33}(Si_{3.65}Al_{0.35})O_{10}(OH)_2$$

The amount of decrease in tetrahedral Al is equal to the increase in octahedral Al. Thus, assuming this is a true weathering sequence, as the Mg is stripped from the octahedral sheet it is preferentially replaced by Al removed from the tetrahedral sheet. Under low-temperature, slightly acid conditions Al strongly prefers the octahedral coordination. Fe also replaces some of the Mg but presumably is derived from another mineral.

The difference in the composition of these two size fractions is similar to the difference between the two types of clay vermiculite described by Barshad and Kishk. The two sets of data confirm the idea that clay vermiculites developed by mild leaching action of pre-existing sheet structures tend to inherit much of their octahedral and tetrahedral character. Clay vermiculites formed by relatively intense weathering and by diagenetic alterations and in approximate equilibrium with their soil environment will have little if any tetrahedral Al; the octahedral sheet can be quite variable in composition and depends on the availability of Al, Fe, and Mg.

Quite often Al, Fe, and Mg hydroxides partially fill the interlayer position of the "derived" vermiculites and decrease their exchange capacity and their ability to contract completely to 10 Å when heated or when treated with a potassium solution. This material can usually be removed by treating the clay with a solution of sodium citrate (Tamura, 1958). As the content of hydroxy interlayer material increases, the expandable clay tends to assume the character of a chlorite. Thus, in the weathering of a mica or illite it is not uncommon to form discrete vermiculite-like, beidellite-like, montmorillonite-like and chlorite-like layers. These various layers can occur as discrete packets or interstratified in a wide variety of proportions.

Roth et al. (1969) have demonstrated that clay vermiculites commonly have a surface coating of positively charged Fe_2O_3 (approximately 10% Al_2O_3 is also present) which causes a decrease in C.E.C. This material can be removed (deferration) by reducing free Fe_2O_3 with Na_2SO_4 in the presence of Na citrate and $NaHCO_3$.

Much of the derived expanded clay, even that which resembles montmorillonite (holds two layers of ethylene glycol), will contract to 10 Å when exposed to a potassium solution. Weaver (1958) has shown that these clays can obtain sufficient potassium from sea water and readily contract to 10 Å. Vermiculite and mixed-layer biotite-vermiculites are rare in marine sedimentary rocks. Weaver (1958) was unable to find any expandable clays in marine sediments that would contract to 10 Å when treated with potassium. A few continental shales contained expanded clays that would contract to 10Å when saturated with potassium. Most vermiculites derived from micas and illites have high enough charge so that when deposited in sea water they extract potassium and eventually revert to micas and illites. Some layers may be weathered to such an extent that they do not have sufficient charge to afford contraction and mixed-layer illite-montmorillonites form.

So-called mixed-layer chlorite-vermiculites are common in marine sedimentary rocks, but it appears that in most, if not all, instances the vermiculite layers will not contract when saturated with potassium and the expanded layers are probably some form of smectite. These clays probably formed from volcanic material, montmorillonite or chlorite, rather than from the degradation of micas and illites.

Hydrothermal studies by Roy and Romo (1957) indicate that macroscopic vermiculite cannot form above 200–300°C. At these temperatures some of the octahedral ions migrate into the interlayer positions and a "chlorite-like" phase is formed.

Chapter 8

MIXED-LAYER CLAY MINERALS

Mixed-layer or interstratified clay minerals are those in which individual crystals are composed of unit cells or basic unit layers of two or more types. It is quite probable that the great majority of clay minerals are composed of interstratified layers of differing composition. In most instances these differences are not detected by routine methods of analysis. From a practical standpoint, clay minerals are classified as interstratified when the layers are sufficiently different in character and sufficiently abundant that the presence of the two or more layer types can be established by X-ray analysis.

The layers may be regularly or randomly interstratified. The latter are by far the more common and are probably the second most abundant clay mineral species, following illite (which in most cases is a mixed-layer clay). More regularly interstratified clay minerals have a definite periodicity and some are given specific mineral names. The randomly interstratified clay minerals are described in terms of the type and proportion of the two or more types of layers. Many of them exhibit some degree of regular interlayering.

Illite-montmorillonite

The most common type of mixed-layer clay is composed of expanded, water-bearing layers and contracted, non-water-bearing layers (i.e., illite-montmorillonite, chlorite-vermiculite, chlorite-montmorillonite). Most of these clays form by the partial leaching of K or Mg $(OH)_2$ from between illite or chlorite layers and by the incomplete adsorption of K or $Mg(OH)_2$ on montmorillonite- or vermiculite-like layers. They most commonly form during weathering or after burial but are frequently of hydrothermal origin.

Regular mixed-layer illite-montmorillonites (Table XLV) composed of alternating mica-like and montmorillonite-like layers have been described by several authors; however, Brown and Weir (1963) proposed that those which had previously been described under other specific names, such as allevardite, were forms of the mineral rectorite. Rectorite is composed of high-charge (mica-like) layers with fixed interlayer cations alternating in a regular manner with low-charge (montmorillonite-like) layers that have exchangeable cations capable of hydration. The expandable layers are both beidellitic and montmorillonitic (Kodama, 1966). Although the mica-like layers generally contain

TABLE XLV

Chemical analyses and structural formulas of rectorite-like clays

	1	2	3	4	5
SiO_2	53.7	54.11	55.2	49.93	54.15
Al_2O_3	38.2	40.38	38.7	25.97	36.12
Fe_2O_3	0.77	0.15	0.59	2.01	2.19
FeO	0.65	–	–	0.30	–
MgO	0.42	0.78	–	2.77	0.65
CaO	1.19	0.52	0.33	1.11	1.50
Na_2O	2.68	3.87	4.40	0.09	–
K_2O	1.31	0.29	0.11	3.46	5.54
TiO_2	0.36	0.01	–	0.42	0.06
H_2O^+	–	–	–	7.45	–
H_2O^-	–	–	–	6.49	–
Total	99.46[a,b]	100.24[a,c]	99.33[a]	100.00	100.47[a,d]

		1	2	3	4	5
Octahedral						
	Al	1.93	2.01	2.01	1.64	1.88
	Fe^{3+}	0.04	0.01	0.03	0.11	0.10
	Fe^{2+}	0.03	–	–	0.02	–
	Mg	0.04	0.07	–	0.29	0.02
	Σ	2.04	2.09	2.04	2.06	2.00
Tetrahedral						
	Al	0.77	0.80	0.71	0.50	0.70
	Si	3.23	3.20	3.29	3.50	3.30
Interlayer						
	Ca and Mg	0.16	0.07	0.04	0.17	0.12
	Na	0.31	0.44	0.51	0.01	–
	K	0.10	0.02	0.01	0.31	0.45
Charge						
	octahedral	+0.05	+0.20	+0.12	–0.13	–0.02
	Total	–0.72	–0.60	–0.59	–0.63	–0.72

1. Hénin et al. (1954): allevardite, Allevard, France.
2. Kodama (1966): Fort Sandeman district, Baluchistan, Pakistan.
3. Brown and Weir (1963): Na-saturated sample from Garland Co., Ark., U.S.A.
4. Rateyev et al. (1969): K-rectorite, weathered clay, Samarskaya Luka, U.S.S.R.
5. Cole (1966): rectorite-like clay formed from weathered feldspar, Surges Bay, Tasmania.

[a]ignited weight basis; [b]includes 0.18% P_2O_5; [c]includes 0.13% SrO; [d]includes 0.26% P_2O_5.

interlayer Na (paragonite), for some samples (Cole, 1966; Rateev et al., 1969) K is the predominant cation.

The structural formulas, which are presumably an average of at least two or more types of layers, for the Na rectorites indicate the negative charge originates in the tetrahedral sheet and that the octahedral sheets have a slight positive charge. The octahedral positions are almost entirely filled with Al. This suggests these minerals are of hydrothermal origin or have formed from the leaching of paragonitic material. The homogeneity of the octahedral sheet may favor the development of regularly alternating layers.

Detailed Fourier transform studies indicate some of these long-spacing (25 Å–30 Å) clays are not ideally regular. A sample (No.5) described by Cole (1966) contains 72% rectorite-like layers (AB), 18% single mica layers (A), and 10% double mica layers (AA) for an average of 64% mica layers and 36% expandable layers.

TABLE XLVI

Chemical analyses of some mixed-layer illite-montmorillonite clays

	1	2	3	4	5	6	7	8	9	10
SiO_2	51.5	53.6	51.5	52.44	56.11	49.0	54.64	54.9	51.69	60.01
Al_2O_3	28.5	21.7	22.8	26.38	24.38	22.2	26.29	18.1	34.50	23.06
Fe_2O_3	0.61	2.81	3.89	0.31	1.74	6.10	1.51	7.30	0.19	2.10
FeO	0.91	1.24	2.07	0.00	0.58	0.72	0.38	1.19	0.00	0.73
MgO	3.0	3.4	2.0	3.57	3.30	1.3	3.03	2.7	1.56	4.40
CaO	0.05	0.03	0.02	0.66	1.49	0.05	1.40	0.05	0.16	0.22
Na_2O	0.10	0.08	0.11	0.16	0.00	–	0.12	0.04	1.03	0.17
K_2O	9.07	7.08	7.33	7.85	5.13	5.16	5.94	4.41	3.56	2.65
TiO_2	0.77	0.82	0.75	0.44	0.38	1.45	–	0.81	0.68	0.22
H_2O^+	5.5	7.7	6.0	4.78	7.17	10.4	6.61	6.5	5.96	6.91
H_2O^-	0.7	2.2	2.5	3.07	–	3.8	–	2.9	–	–
Total	100.71	100.66	98.97	99.76	100.28	100.18	99.92	98.90	99.33	100.47

1. Hower and Mowatt (1966): Silurian Interlake well core, Williston Basin, Mont., U.S.A.
2. Hower and Mowatt (1966): Cambrian, Eau Claire, Illinois Basin, U.S.A.
3. Hower and Mowatt (1966): Upper Cambrian, Gros Ventre, Wind River Canyon, Wyo., U.S.A.
4. Weaver (1953): Ordovician, Salona, Pa., U.S.A.; analyst J.A. Maxwell.
5. Byström (1956): Ordovician, Sweden.
6. Hower and Mowatt (1966): Upper Cretaceous, Two Medicine, Bowman's Corners, Mont., U.S.A.
7. Bradley (1945): bravasite.
8. Hower and Mowatt (1966): Upper Cretaceous Colorado, Bozeman Pass, Mont., U.S.A.
9. Heystek (1955): hydrothermal alteration of illite, South Africa; analyst D. Sampson.
10. Byström (1956): Ordovician, Sweden.

Rateyev et al. (1969) reported on a similar sample (No.4) which averaged 57% contracted and 43% expandable layers.

Random mixed-layer illite-montmorillonite is by far the most abundant of the mixed-layer clays. These two types of layers occur intergrown in all proportions. Tables XLVI and XLVII contain chemical data for a series of mixed-layer illite-montmorillonites with expanded layers ranging from less than 10% to 60%. When expandable layers comprise more than 60% and less than 10% it is difficult to accurately determine the number of illitic layers. Further, Hower (1967) has shown that many of these clays are partially regularly and partially randomly interstratified. This further complicates interpretation of the X-ray data.

TABLE XLVII

Structural formulas of some mixed-layer illite-montmorillonite clays

	1	2	3	4	5	6	7	8	9	10
Octahedral										
Al	1.66	1.45	1.47	1.64	1.57	1.49	1.62	1.27	1.96	1.50
Fe^{3+}	0.03	0.16	0.20	0.02	0.09	0.33	0.07	0.38	0.01	0.10
Fe^{2+}	0.05	0.07	0.10	0.00	0.03	0.04	0.02	0.07	0.00	0.04
Mg	0.30	0.34	0.21	0.36	0.30	0.14	0.30	0.27	0.15	0.38
Σ	2.04	2.02	2.00	2.02	1.99	2.00	2.01	1.99	2.12	2.02
Tetrahedral										
Al	0.57	0.31	0.40	0.45	0.32	0.42	0.42	0.20	0.66	0.17
Si	3.43	3.69	3.60	3.55	3.68	3.58	3.55	3.80	3.34	3.83
Interlayer										
Ca and Mg	0.02	0.05	0.07	0.10	0.09	0.13	0.20	0.17	0.02	0.15
Na	0.01	0.01	0.01	0.02	0.00	–	0.02	–	0.13	0.01
K	0.77	0.62	0.65	0.64	0.43	0.48	0.50	0.39	0.29	0.22
Charge										
octahedral	0.23	0.35	0.33	0.30	0.36	0.18	0.29	0.37	+0.21	0.36
tetrahedral	0.57	0.31	0.40	0.45	0.32	0.42	0.42	0.20	0.66	0.17
total	0.80	0.66	0.73	0.75	0.68	0.62	0.71	0.57	0.45	0.53
C.E.C. (mequiv./100 g)	12	20	24	27	46	37	–	50	48	77
Percent expanded layers	<10	10	15	20	30	33	40	45	50	60

For legend, see Table XLVI.

TABLE XLVIII

Correlation coefficients for the oxides in fourteen mixed-layer illite-montmorillonites

Variable											
No. name	1	2	3	4	5	6	7	8	9	10	
1 SiO_2	1.000										
2 Al_2O_3	−0.491	1.000									
3 Fe_2O_3	0.375	−0.792	1.000								
4 FeO	0.699	−0.254	−0.037	1.000							
5 MgO	0.461	−0.592	0.288	0.213	1.000						
6 CaO	0.026	−0.389	0.228	0.333	−0.090	1.000					
7 Na_2O	−0.248	0.328	−0.121	−0.072	−0.316	−0.299	1.000				
8 K_2O	−0.398	0.320	−0.622	−0.249	0.051	−0.057	−0.280	1.000			
9 TiO_2	−0.047	0.196	0.181	−0.234	−0.518	−0.250	0.378	−0.387	1.000		
10 H_2O^+	0.563	−0.470	0.738	−0.021	0.289	−0.093	−0.303	−0.471	0.296	1.00	
11 H_2O^-	−0.599	−0.087	−0.182	−0.140	−0.172	0.439	0.009	0.214	−0.319	−0.62	

TABLE XLIX

Correlation coefficients for the cations in fourteen mixed-layer illite-montmorillonites

Variable									
No. name	1	2	3	4	5	6	7	8	9
1 Oct. Al	1.000								
2 Oct. Fe^{3+}	−0.847	1.000							
3 Oct. Fe^{2+}	−0.105	−0.101	1.000						
4 Oct. Mg	−0.557	0.168	0.076	1.000					
5 Tet. Al	0.841	−0.592	−0.524	−0.485	1.000				
6 Tet. Si	−0.842	0.602	0.503	0.486	−0.998	1.000			
7 Int. Ca	−0.155	0.114	0.294	−0.289	−0.206	0.186	1.000		
8 Int. Na	0.366	−0.044	−0.252	−0.548	0.436	−0.441	−0.352	1.000	
9 Int. K	0.277	−0.529	−0.336	0.069	0.394	−0.406	0.106	−0.116	1.000

Correlation coefficients (14 samples) for the oxides and the cations in their various structural positions are given in Tables XLVIII and XLIX, respectively. In addition to the forced correlations, there are several others of interest. Al_2O_3 is negatively correlated with Fe_2O_3 and MgO. Tetrahedral Al and octahedral Al are positively correlated. Tetrahedral Al is negatively correlated with all other octahedral cations and tetrahedral Si is positively correlated with them. This presumably indicates the illitic layers have high tetrahedral charge and the montmorillonitic layers relatively high octahedral charge.

The negative relation between K_2O and Fe_2O_3 also occurs in the illites and has been discussed. Fe_2O_3 is positively correlated with H_2O^+ and presumably negatively with layer charge. H_2O^+ (greater than 110°C) and H_2O^- (less than 110°C) are inversely related. These correlations suggest some of the H_2O^+ may be trapped interlayer water not easily released (at 110°C) from the clays with a high proportion of contracted layers or that chloritic layers increase as the proportion of expandable layers decrease.

Hower and Mowatt (1966) have made plots which show K (+Na) increases as the proportion of 10 Å layers increase and as total layer charge increases. There is also a

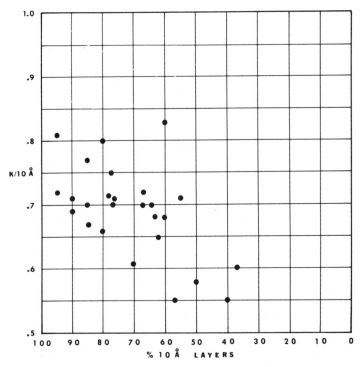

Fig.18. The relation of the percentage of contracted layers (10 Å) to the amount of K per 10 Å layer for twenty-six mixed-layer illites-montmorillonites.

vague positive relation between K and tetrahedral charge. The first plot extrapolates to a value of 0.75 $(K + Na)_n$ for 100% contracted layers. However, when a plot is made of percent 10 Å layers versus K_2O the linear relation extrapolates to about 9% K_2O for 100% 10 Å layers. Hower (1967) has suggested that mixed-layer illite-montmorillonites of a given bulk composition vary in their ability to expand depending on the amount of ordering. Randomly interstratified 2:1 units result in significantly greater expandability than an ordered structure of the same composition.

When all the K is assigned to the contracted 10 Å layers and a plot made of the amount of K per 10 Å layer versus percent 10 Å layers (Fig.18), a distinct positive relation is apparent. The data indicate that when there are no expanded layers the clay contains 0.8 K per $O_{10}(OH)_2$; the amount of K systematically decreases and for 40% contracted layers the concentration is 0.55 K per $O_{10}(OH)_2$. If some K is present in the expanded layers the amount of K in the contracted layers would be even lower. These data may indicate the calculated proportion of contracted layers is high or, on the basis of Hower's (1967) reasoning, the amount of ordered interlayering increases with a decrease in the proportion of contracted 10 Å layers (less average charge per layer is required to contract ordered interlayers than random interlayers). Another possibility is that approximately 20% of the contracted layers are chlorite. It is difficult to detect less than 40% illite layers interlayered with montmorillonite and it is probably equally difficult to detect 20% or so chloritic layers (Weaver and Beck, 1971a).

Weaver (1965) has attempted to calculate the structural formula of the two types of layers in a mixed-layer (75:25) Ordovician K-bentonite by assuming all of the K is in the contracted layers and the expandable layers contain no tetrahedral Al.

$$\text{Illite: } \overset{K_{0.84}}{(Al_{1.67}Mg_{0.32})} (Al_{0.60}Si_{3.40})O_{10}(OH)_2$$

$$\text{Montmorillonite: } \overset{Ca_{0.19}Na_{0.09}}{(Al_{1.54}Mg_{0.46})} (Si_{4.00})O_{10}(OH)_2$$

The formulas are reasonable but somewhat oversimplified.

As has been noted (Weaver, 1953; Ormsby and Sand, 1954; Byström, 1956; Hower and Mowatt, 1966), the cation exchange capacity is closely related to the proportion of expanded layers. If all the exchange capacity is assigned to the montmorillonite-like layers, values ranging from 96 to 144 mequiv./100 g are obtained which are within the range of values obtained for pure montmorillonites.

Mixed-layer clays form by the alteration of pre-existing micas and illites, by hydrothermal action, by the alteration of volcanic glass, and by diagenetic alteration of montmorillonite. During continental weathering K is leached from micas and illites and mixed-layer clays are formed. When these clays are carried to the sea, they may adsorb K and revert partially or entirely to illite. If they remain in a continental environment where K may not be available, the expanded layers can persist until K becomes available. Expanded layers formed in this manner have a higher layer charge than those formed in a marine environment.

Many of these weathered micas and illites are completely leached of K and have the swelling characteristics of a montmorillonite. This material is usually leached to such an extent that the layer charge of some of the layers is sufficiently lowered so that they will no longer contract to 10 Å. When these clays are exposed to sea water or K from any source, only a portion of their layers will contract and a mixed-layer clay is formed.

Diagenetic modification of expandable clay during burial is an important source of mixed-layer illite-montmorillonite. With increasing depth of burial and increasing temperature the proportion of contracted 10 Å layers systematically increases. From about $50°C-100°C$ the contracted layers are distributed randomly. At higher temperatures only a few additional layers are contracted but the interlayering becomes more ordered (Perry and Hower, 1970; Weaver and Beck, 1971a). The "final" product, 7:3 to 8:2, is relatively stable and persists until temperatures on the order of $200°C-220°C$ are reached.

K is obtained from associated K-feldspars and micas. The layer charge is increased by the reduction of iron in the octahedral sheet and incorporation of Al, entering through the ditrigonal holes in the basal oxygen plane, into the tetrahedral sheets (Weaver and Beck, 1971a; Pollard, 1971). Weaver and Beck have presented evidence that indicates mixed-layer clays formed in this manner contain 20-30% chloritic layers and are actually mixed-layer illite-chlorite-montmorillonite clays.

Chlorite-montmorillonite

The second most abundant type of mixed-layer clay is that composed of the intergrowth of chlorite and 2:1 water bearing layers. These latter layers may be montmorillonite-like (two layers of ethylene glycol) or vermiculite-like (one layer of ethylene glycol). The chlorite and expanded layers occur in all proportions; however, only samples in which the ratio is approximately 1:1 have been analyzed chemically. Such clays are regularly interstratified and the name corrensite or swelling chlorite has been applied to some of them. MacEwan et al. (1961) have suggested that the name corrensite be restricted to the type mineral described by Lippman (1954). In this mineral, the expandable layers appear to contain a sufficient amount of brucite material to prevent the layers from contracting to 10 Å when heated. The mixed-layer clays described by Bradley and Weaver (1956), Earley et al. (1956), and others readily contract to 10 Å. Most of these clays for which chemical analyses are available will adsorb two layers of ethylene glycol and probably should be called chlorite-montmorillonites rather than chlorite-vermiculites. Most of the clays of this type are trioctahedral although Sudo and Kodama (1957) described a sample in which the octahedral sheet of the 2:1 layer is dioctahedral and the hydroxide sheet is trioctahedral. Shimoda (1969) described a regular mixed-layer di/trioctahedral chlorite-montmoril-

TABLE L

Chemical analyses of some mixed-layer chlorite-montmorillonites

	Trioctahedral					Dioctahedral	
	1	2	3	4	5	6	7
SiO_2	41.2	43.1	37.2	36.12	33.44	39.94	42.14
Al_2O_3	12.1	16.6	15.5	13.67	14.33	33.17	37.38
Fe_2O_3	1.74	6.32	6.7	2.14	1.66	1.34	0.30
FeO	0.39	–	–	6.85	3.68	0.18	–
MgO	22.0	17.65	18.9	25.16	22.44	6.44	0.08
CaO	1.4	0.94	1.0	1.70	2.83	1.30	1.65
Na_2O	0.07	0.52	0.2	0.37	0.16	0.52	0.15
K_2O	0.22	2.72	1.4	0.10	tr	0.24	1.40
TiO_2	0.04	0.73	0.4	0.23	–	0.74	–
H_2O^-	6.80	–	–	4.81	5.50	4.39	6.16
H_2O^+	15.4	7.40	18.4	9.25	8.16	11.64	11.22
Total	101.36	100.48[a]	99.7	100.40	92.36[b]	100.67[c]	100.48
C.E.C. (mequiv./100 g)	40			24	47		

1. Bradley and Weaver (1956): Upper Mississippian limestone, Juniper Canyon, Colo., U.S.A., contains approximately 20% quartz.
2. Earley et al. (1956): Permian, Yates siltstone, Ward Co., Texas, U.S.A., contains dolomite, mica and organic material; analyst B.B. Osthaus.
3. Peterson (1961): Upper Mississippian dolomite, Sparta, Tenn., U.S.A., contains illite and amorphous Fe_2O_3; analyst J. Ito.
4. Alietti (1957): hydrothermal with serpentine, Taro Valley, northern Appennines, contains serpentine; analyst A. Alietti.
5. Mongiorgi and Morandi (1970): secondary hydrothermal alteration of Al-saponite; breccia associated with diabase, near Rossena, Italy.
6. Sudo and Kodama (1957): hydrothermal, altered halo around pyrite ore body in volcanic rocks, contains pyrite.
7. Shimoda (1969): alteration mineral of Tertiary tuff and tuffaceous sediments around gold-silver-quartz veins.

[a] includes 0.03% MnO, 0.26% P_2O_5, 0.81% CO_2, 3.40% C; [b] includes 0.69% S, 0.08% P_2O_5; [c] includes 0.16% MnO; analysis corrected for 10% datolite.

lonite (given the specific name *tosudite*). Chemical analyses are given in Table L. All of the samples described have some impurities. This, coupled with the inability to properly proportion the octahedral sheets, makes the calculated structural formula of little value. By attempting to balance the layer charges the following compositions have been calculated for the tetrahedral sheets:

Sample No. 1 $(Si_{2.73}Al_{1.27})O_{10}(OH)_2$
Sample No. 2 $(Si_{3.42}Al_{0.58})O_{10}(OH)_2$
Sample No. 3 $(Si_{3.32}Al_{0.68})O_{10}(OH)_2$

There is nothing to indicate whether there is any difference between the tetrahedral and octrahedral sheets of the chlorite layer and the montmorillonite layer. In the trioctahedral clays, Mg is the dominant (60–90%) octahedrally coordinated cation. In the dioctahedral clay, Al is the dominant octahedral cation.

These clays occur in limestones, dolomites, evaporites, shales, siltstones, and hydrothermal deposits. All the sedimentary material appears to have a diagenetic origin. Although the physical environments vary, the chemical environments should be similar. Saline or even super-saline conditions are implied by the presence of evaporite minerals associated with some of the deposits. In the other deposits it is possible that temporary evaporitic conditions (e.g., tidal flats) existed long enough for brucite to precipitate between the layers of expanded-layer minerals. It appears plausible that the parent material was a montmorillonite-like mineral (probably detrital in most cases).

Although the chlorite and montmorillonite layers occur intergrown in a variety of proportions, the 1:1 regular intergrowth appears to be the most well developed from a crystalline viewpoint and is presumably in equilibrium with the Mg-rich environments in which it forms (Grim et al., 1961; Peterson, 1961). An explanation for the cause of the regularity is given by Grim et al. (1961) in the following statement:

"Heterogeneous equilibrium at moderate pH in the liquid phase apparently requires the presence of both chlorite (basic) and vermiculite (acidic) in the solid. In view of the success of the important Pauling Principle that a complex solid structure tends to become electrostatically neutral in the smallest practical volume, it seems only natural that its extension to neutrality in the acid–base sense is equally valid. For the articulated layer structures, best economy of space is achieved by regularly alternating intergrowth of the two species of layers."

Another explanation might be similar to that proposed by Bassett (1959) for regularly interstratified biotite-vermiculite. On the basis of the composition it is likely that both the interlayer brucite sheet and the octahedral sheet in the 2:1 layer have a positive charge and the tetrahedral sheet has a negative charge. In trioctahedral smectites and vermiculites the apex oxygens in the Al tetrahedra carry much of the negative charge which shortens and strengthens the bonds to the octahedral cations in the 2:1 layer. Once the pH and Mg-Al ion activity of the fluid becomes such that a positively charged brucite sheet develops between two tetrahedral sheets much of the net negative charge in these sheets is directed toward the brucite sheet leaving only the one tetrahedral sheet to satisfy the charge of the 2:1 octahedral sheet. The net negative charge available on the other side of the 2:1 sheet would be relatively small and conceivably even positive. The result would be that the repulsive forces between the basal oxygens of the tetrahedral sheet on the side of the 2:1 sheet away from the brucite interlayer and the basal oxygens of the adjacent sheet would be considerably reduced (perhaps temporarily allowing such a close approach that a brucite sheet could not develop) and the total charge and the asymmetrical charge distribution would not be particularly favorable for the development of a positive brucite sheet.

Additional brucite sheets would tend to develop between unaffected sheets and avoid interlayer spaces adjacent to where a brucite sheet had already been organized. Initially, these brucite sheets would be distributed randomly but as they approached the point where they filled approximately half the interlayer spaces, the charge distribution and minimum free-energy requirements would dictate that the final brucite

TABLE LI

Chemical analyses and structural formulas of some regular mixed-layer clays

	1	2	3		1	2**	3
SiO_2	54.31	49.42	22.0	Octahedral			
Al_2O_3	1.90	4.97	27.6	Al	–	–	2.12
Fe_2O_3	–	4.81	4.7	Fe^{3+}	–	0.20	0.38
FeO	2.95	1.54	30.2	Fe^{2+}	0.17	0.09	2.74
MgO	27.44	28.56	4.7	Mg	2.86	2.80	0.76
CaO	1.01	0.60	–		3.03	3.09	6.00
Na_2O	0.32	0.38	–	Tetrahedral			
K_2O	0.86	0.04	–	Al	0.16	0.45	1.41
TiO_2	–	0.21	–	Si	3.80	3.48	2.39
H_2O^+	9.48	8.35	10.6	Fe^{3+}	–	0.07	–
H_2O^-	2.59	–	0.10	Interlayer			
Total	100.89	98.91*	99.9	Mg	–	0.006	–
				Ca/2	0.07	0.02	–
				K	0.07	–	–
				Na	0.04	0.01	–

1. Allieti (1956): mixed-layer talc-saponite, Monte Chiaro, Italy.
2. Veniale and Van der Marel (1968): mixed-layer talc-saponite, Ferriere, Nure Valley, Italy.
3. Brindley and Gillery (1954); Brindley (1961): mixed-layer serpentine-chlorite, Cornwall, England; analyst R.F. Youell.

*analysis before correction for estimated 10% chrysotile; also contained 0.03% P_2O_5 and trace MnO.
**after correction for 10% chrysotile.

sheets tend to fill those remaining interlayer spaces which most nearly afford a regular alternation of brucite interlayers (relatively large positive charge) and cation-water interlayers (relatively small positive charge).

Other mixed-layer clay minerals

There is a variety of other types of mixed-layer clays, but identifications and chemical analyses are not numerous. A regular mixed-layer talc-saponite from Italy was described by Allietti (1956). Another talc-saponite also from Italy was described by Veniale and Van der Marel (1968). They proposed that this regular mixed-layer talc-saponite formed by the hydrothermal serpentinization of the parent ultramafic rock. The clay then accumulated as a residual mineral inside the surface weathering products. Interstratified talc-saponite has also been reported as having a sedimentary origin (Guenot, 1970). An iron-bearing chlorite (classified as *daphnite*) from Cornwall, England, was said by Brindley and Gillery (1954) to have a mixed-layer serpentine-chlorite structure. Chemical data for these mixed-layer clays are given in Table LI.

Chapter 9

ATTAPULGITE AND PALYGORSKITE

Attapulgite and palygorskite have a fibrous texture and a chain structure. The structure proposed by Bradley (1940) is that of a 2:1 layer structure with five octahedral positions (four filled); four Si tetrahedra occur on either side the octahedral sheet with their apices directed towards the octahedral sheet. These structural units alternate in a checkerboard pattern leaving a series of channels between the structural units. These channels contain water molecules.

Huggins et al. (1962) found that the palygorskite samples they studied (9 through 13, Table LII) contained long fibers; whereas, the structurally identical clay from Attapulgus, Georgia, consisted of short (~1μ) fibers. Although most of the Georgia commercial attapulgite consists of short fibers, the long-fiber variety is relatively common. The latter variety occurs in layers and patches within beds consisting of short fibers. The two lengths are quite distinctive and although they are structurally similar, they probably differ chemically. The available data indicate that the short variety has more iron than the long variety. The fibrous clay in soils and "normal" sediments is most commonly the short variety. It would seem desirable to retain both names as future studies will probably indicate they are two distinct fiberous species.

Christ et al. (1969) found that X-ray diffraction powder data for palygorskite samples show both orthorhombic and monoclinic structures and suggested that the variations in symmetry reflect variations in chemical composition. The present data suggest the most likely difference is octahedral Fe.

Only a few analyses have been made of palygorskite but it is enough to indicate that the composition is probably as variable as that of the 2:1 minerals. Some of the variations are due to the presence of montmorillonite which is commonly intimately associated with the attapulgite and is difficult to remove.

Chemical analyses of 15 palygorskite samples are listed in Table LII and structural formulas in Table LIII. These formulas were calculated on the basis of 21 oxygens per half-unit-cell of the dehydrated clay. Histograms are shown in Fig.19. There are no well-developed model values for any of the structural positions (Fig.19); however, the total octahedral cations have a normal distribution with a mode at 3.9. The H^+ content of the octahedral sheet is so speculative that it is impossible to make a reasonable calculation of the layer charge. The tetrahedral aluminum ranges from 0.01 to 0.69 per eight tetrahedral positions which is similar to the range for low-aluminum montmorillonites. Octahedrally coordinated aluminum and total Al_2O_3 is less than is

TABLE LII

Chemical composition of palygorskite samples

	1	2	3	4	5	6	7	8	9	10	11	12	13	14	15
SiO_2	61.60	54.71	53.64	52.35	52.6	54.25	52.18	50.65	57.54	57.19	55.71	55.12	51.17	55.86	57.17
Al_2O_3	6.82	13.48	8.76	15.44	12.6	13.11	18.32	11.97	12.26	12.13	12.83	15.70	13.73	10.54	13.25
Fe_2O_3	0.87	2.10	3.36	2.12	3.8	–	–	7.45	0.21	1.51	1.39	1.60	1.55	3.23	6.25
FeO	–	–	0.23	–	0.8	–	–	–	–	–	–	–	0.31	–	–
TiO_2	–	–	0.60	–	–	–	–	0.20	–	–	–	–	–	0.47	0.36
CaO	0.67	2.79	2.02	0.14	2.2	0.31	0.59	0.14	0.08	0.36	0.18	0.41	2.89	1.56	0.69
MgO	14.22	5.44	9.05	6.60	8.4	12.04	8.9	7.75	10.22	9.29	9.20	6.14	6.40	9.20	7.29
Na_2O	–	–	–	–	–	–	–	–	–	0.15	0.45	–	–	0.68	–
K_2O	–	–	–	–	–	–	–	–	0.18	0.07	0.11	–	–	0.05	–
H_2O^+	–	12.63	10.89	12.00	10.6	13.36	12.04	10.56	–	–	–	13.52	13.24	9.13	15.04
H_2O^-	14.16	8.65	9.12	10.32	9.0	6.50	8.46	9.72	–	–	–	7.08	10.29	8.71	–
Total	98.34	99.80	97.67*	98.97	100.0	99.57	99.78	99.10**	100.24	100.24	100.68	100.96	99.57	99.58	99.85
C.E.C. (mequiv./100 g)	20	21		36				28							

1. Heystek and Schmidt (1954): Dornboom, South Africa.
2. Caillère (1934): Taodeni, Algeria.
3. Bradley (1940): Attapulgus, Ga., U.S.A., short.
4. Stephen (1954): Bakkasetter, Shetland, long.
5. Caillère (1951): Tafraout, Morocco.
6. Fersmann (1913): Dogniaska, U.S.S.R.
7. Fersmann (1913): Permsk, U.S.S.R.
8. Ovcharenko (1964): Cherkassy, Ukrain, U.S.S.R.
9. Huggins et al. (1962): Brazil, long.
10. Huggins et al. (1962): Metaline Falls, Wash., U.S.A., long.
11. Huggins et al. (1962): Lemesurier Island, Alaska, long.
12. Huggins et al. (1962): Volhynia, U.S.S.R., long.
13. Huggins et al. (1962): Gor'kiy, U.S.S.R., long.
14. Huggins et al. (1962): Attapulgus, Ga., U.S.A., short.
15. Wiersma (1970): Jordan Valley, Israel, short.

*P_2O_5, K_2O, Na_2O, MnO total 2.40%; **includes 0.1% MnO + 0.56% K_2O + Na_2O.

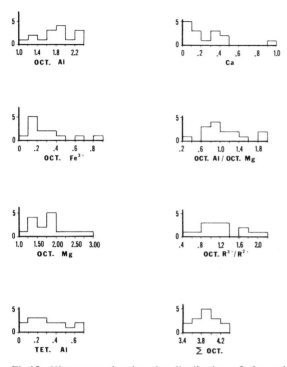

Fig.19. Histograms showing the distribution of the cations of fifteen palygorskite structural formulas.

found in the montmorillonites. The magnesium content of the octahedral sheet and MgO is 2–4 times as abundant as in montmorillonite. The iron contents are similar.

The average Al_2O_3/MgO ratio for 24 illites is 9.6 and for 101 montmorillonites 6.7. Attapulgite values range from 2.5 to 0.48. The ratios of octahedral Al/octahedral Mg are respectively 5.4, 4.3 and 1.8–0.4. Radoslovich (1963b) found that the 2M muscovite structure required a minimum of 1.7 of the three octahedral sites be filled with Al. The Al occurs in the two symmetrically related sites and the larger divalent cation occurs in the distinctive or "unoccupied" site. The lower limit of 1.7 Al is equivalent to 85% of the two symmetrically related or occupied sites being filled in a stable muscovite structure. A similar restriction is reported for the trioctahedral micas where an upper limit of 1.00 ($R^{3+} + R^{4+}$) per three sites was found by Foster (1960).

For 24 illites an average of 1.53 Al per three octahedral sites was found. This is 77% occupancy of the two occupied sites. Total trivalent ions (Al+ Fe^{3+}) averaged 1.76. Thus, most illites have near the minimum or less than the minimum number of Al per three sites required to maintain a stable $2M_1$ muscovite structure.

The average Al per three sites for 101 montmorillonites is 1.49 (50%) and for Al + Fe^{3+}, 1.68. Frequency distribution graphs of data from these two mineral groups

TABLE LIII

Palygorskite structural formulas

	1	2	3	4	5	6	7	8	9	10	11	12	13	14	15
Octahedral															
Al	1.13	2.11	1.29	2.25	1.63	1.74	2.34	1.34	1.92	1.87	1.94	2.32	1.80	1.50	1.73
Fe^{3+}	0.10	0.23	0.37	0.23	0.41			0.81	0.02	0.16	0.14	0.17	0.17	0.33	0.63
Fe^{2+}			0.03		0.09								0.04		
Mg	3.02	1.16	1.96	1.44	1.79	2.53	1.72	1.66	2.09	1.91	1.92	1.28	1.39	1.90	1.45
Σ	4.25	3.50	3.65	3.92	3.92	4.27	4.06	3.81	4.03	3.94	4.00	3.77	3.40	3.73	3.81
Tetrahedral															
Al	0.01	0.16	0.21	0.39	0.49	0.41	0.63	0.69	0.07	0.11	0.19	0.27	0.55	0.23	0.36
Si	7.99	7.84	7.79	7.61	7.51	7.59	7.37	7.31	7.93	7.89	7.81	7.73	7.45	7.77	7.64
Interlayer															
Ca	0.10	0.43	0.32	0.32	0.33	0.04	0.09	0.04	0.02	0.11	0.05	0.13	0.90	0.47	0.20
Octahedral															
Al/Mg	0.37	1.82	0.66	1.57	0.91	0.69	1.36	0.81	0.92	0.98	1.01	1.81	1.30	0.79	1.19
R^{3+}/R^{2+}	0.40	2.02	0.83	1.72	1.08	0.69	1.36	1.29	0.93	1.06	1.08	1.96	1.32	0.99	1.63

indicate that the large majority contain between 1.3 and 1.7 Al per three sites. There are relatively abrupt decreases in frequency at these values (Fig.2,14) suggesting that they are significant boundary values. The minimum value of 1.3 is equivalent to 65% occupancy of the two occupied sites as compared to a minimum of 85% for $2M_1$ muscovite and 75% for 1M biotites.

The structure proposed by Bradley (1940) corresponds to the chemical formula $Mg_5 Si_8 O_{20}(OH)_2(OH_2)_4 \cdot 4H_2O$. As shown in Table LIII only four of the proposed five octahedral sites are occupied. Al fills from 1.13 to 2.34 of these five sites or between 28% and 59% of the four occupied sites. Al + Fe^{3+} values range from 31% to 62%. Divalent cations, largely Mg, occupy from 29% to 76% of the four occupied sites. As an average it would appear that of the four occupied sites approximately half contain trivalent and half divalent cations.

Chemical analyses indicate that in the 2:1 layer structures with predominantly Al and Mg in the octahedral sites, the minimum occupancy for the poorly crystallized

TABLE LIV

Chemical analyses of attapulgite from Attapulgus, Ga. (U.S.A.)

	1	2	3	4	5	6	7
SiO_2	55.04	53.67	55.03	53.64	54.04	53.96	56.00
Al_2O_3	8.96	8.63	10.24	8.76	9.83	8.56	7.98
Fe_2O_3	3.43	3.04	3.53	3.36	3.52	3.10	2.51
FeO				0.23	0.19	0.19	–
MgO	9.52	11.06	10.49	9.05	9.07	10.07	12.40
CaO	1.55	1.55		2.02	1.69	2.01	
Na_2O	0.05	0.42	0.47	0.83	0.08	0.03	
K_2O	0.50	0.35		0.75	0.57	0.39	
TiO_2	0.48	0.49		0.60	0.32	0.24	
H_2O^+	19.60	19.49	10.13	10.89	10.93	11.51	
H_2O^-			9.73	9.12	10.00	9.79	
P_2O_3	0.80	1.30		0.79			
Total	100.46	100.42	99.62	100.04	100.24	99.85	
C.E.C. (mequiv./100 g)	10.60	10.8	21		(99.0)	35.0	

1. Minerals and Chemicals Corp. of America: typical mine-run sample containing montmorillonite.
2. Minerals and Chemicals Corp. of America: 1.0–0.1μ fraction contains 5% montmorillonite, increase in Na_2O and P_2O_5 due to dispersant.
3. Bradley (1940): 0.1–0.05μ fraction.
4. De Lapparent (1935).
5, 6. Kerr et al. (1950): 1–3% impurity.
7. Abdul-Latif and Weaver (1969): purified sample from Minerals and Chemicals Corp. of America.

TABLE LV

Observed water losses (%) compared with those for ideal palygorskite (After Caillère and Hénin, 1961a)

	1	2	3	4	5
Zeolitic water, below 200°C	10.8	10.7	8.0	9.8	8.60
Bound water, 250–400°C	3.6	2.7	4.0	4.0	8.60
Hydroxyl water, above 400°C	4.2	6.2	6.2	6.5	2.15
Total water	18.6	19.6	18.2	20.3	19.35

1. Attapulgus, Ga., U.S.A.
2. Taodeni, Algeria.
3. Tafraout, Morocco.
4. Meyssonial en Mercoeur, Upper Loire, France.
5. Ideal palygorskite, $(Si_8)(Mg_5)O_{20}(OH)_2(OH_2)_4 4H_2O$.

clay minerals is 65% and the minimum for well-crystallized muscovite 85%. In the 1:1 layer silicates occupancy values less than 85% are rare. Celadonite (and some glauconites) appears to be the only layer silicate containing approximately half trivalent and half divalent cations in the occupied octahedral sites. However, in these minerals, the trivalent cation is Fe^{3+} rather than Al and the equivalent Al member has not been found or synthesized (Yoder, 1959).

Radoslovich (1963b) suggests that as the unsatisfied charge shifts from the tetrahedral to the octahedral sheet there is an increase in anion-anion repulsion in the latter sheet which leads to long-shared edges if Fe^{3+} is the trivalent cation. "If Al is substituted for Fe^{3+} then the average cation-oxygen bonds are correspondingly shortened, and the octahedral cations brought closer together — in fact unduly close."

It would seem that when the occupied octahedral sites are more than 65% occupied by Al or by Mg the layer can adjust to compensate for the internal strain and can grow in two dimensions. The minerals which form larger sheets generally have a larger proportion of their occupied sites filled with Al or Mg than the smaller, clay minerals. The occupancy value can be less than 65% when the smaller Fe^{3+} is substituted for Al. When these conditions are not satisfied, the internal strain is such that growth is in only one direction. The width of the sheet is restricted to five octahedral sites. Sufficient strain accumulates within this five-site interval that the silica tetrahedral sheet is forced to invert to accommodate the strain.

Table LIV contains a list of seven analyses of attapulgite from Attapulgus, Georgia. These values are reasonably similar when compared to the seven analyses of Fithian illite. The Al_2O_3/MgO ratio is relatively constant, varying from 0.64 to 1.08.

The structure proposed by Bradley (1940) has three forms of water: a zeolitic water, bound water (at the edges of the octahedral sheet), and structural hydroxyls. Using thermogravimetric curves, Caillère and Hénin (1961a) attempted to measure the amount of these three types of water for several attapulgites (Table LV). The amount of bound water and hydroxyls differs considerably from that calculated on the basis of the ideal structure (column 5) and suggests there are more structural hydroxyls than proposed for the ideal structure.

Cation exchange capacity values range from 10 to 99 mequiv./100 g. These higher values are presumably in error or a result of the presence of large amounts of montmorillonite. The exchange capacity is presumably due to deficits of charge in the structure, but analyses are not sufficiently accurate to prove this. Exchange values of 5–30 mequiv./100 g seem to be most characteristic of the relatively pure samples. CaO, Na_2O, K_2O are all usually reported in analyses of attapulgite and at least some of this material, along with some of the magnesium, is probably present as exchangeable cations; however, in most samples these cations are more abundant than can be accounted for by the cation exchange values. CaO is usually most abundant being present in amounts ranging from 1 to 2%. Examination of hundreds of samples from Georgia indicate illite and calcite are present in most samples. The illite and calcite would account for much of the K_2O and some of the CaO. Loughnan (1959) acid-leached an attapulgite from Ipswich and lowered the CaO content from 5.1% to 0.9%. No CO_2 was found in the acid treated sample. Because the acid treatment would have removed any exchangeable Ca, it is likely that some of the Ca is an integral part of the structure.

A plot of Ca versus Σ octahedral cations shows a well-developed inverse relation (Fig.20). This could indicate that exchangeable Ca increases as the octahedral cation population decreases and octahedral charge increases; however, it appears more likely that the relation indicates some Ca is in the octahedral sheet. The abundance of large cations in the octahedral sites of attapulgite introduces sufficient strain to make it not unlikely that some Ca could be accommodated. As Wiersma (1970) has published a thorough review of the literature, only a sampling of available references will be mentioned. Although attapulgite is relatively rare, it forms in a variety of ways. Caillère (1951), Stephen (1954) and Bonatti and Joensuu (1968) describe hydrothermal origins, the latter under marine conditions. Muir (1951), Millot (1953), Grim (1953), Barshad et al. (1956), Loughnan (1959), Rateev (1963) and Weaver and Beck (1971b) have indicated that it can form in lagoons, playa lakes or evaporatic basins. Rogers et al. (1954) have suggested a lacustrine origin for the Ipswich materials. Attapulgite is commonly associated with or is a component of calcium carbonates, particularly dolomite. Much of the dolomite is evaporitic or supratidal in origin. Arid conditions appear to be essential for the formation of sedimentary attapulgite. These conditions presumably provide the high Mg concentration and the high pH necessary for its formation.

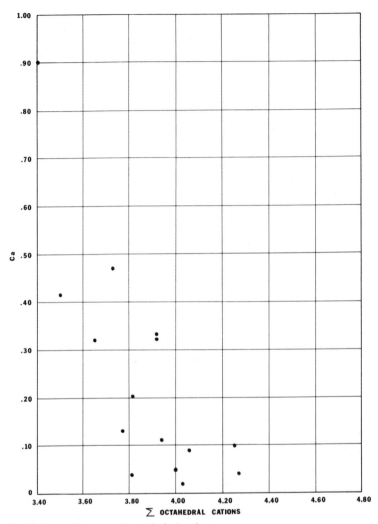

Fig. 20. Plot of Ca versus Σ octahedral cations.

Arid climatic conditions also favor the formation of attapulgite in desert soils. Van den Heuvel (1964) described a calcareous soil from New Mexico which contained abundant attapulgite and sepiolite. The evidence suggests that the attapulgite and sepiolite were formed during soil development. Elgabaly (1962) described attapulgite in the desert soils of Egypt and Al Raivi et al. (1969) in the arid regions of Iraq.

Attapulgite is invariably associated with montmorillonite and in some situations is believed to weather to montmorillonite (Barshad et al., 1956) and by others to be transformed from montmorillonite by the addition of magnesium (Loughnan, 1959). It is not certain that either transformation occurs, although under hydrothermal conditions attapulgite alters to montmorillonite (Mumpton and Roy, 1958).

Chapter 10

SEPIOLITE

Sepiolite is a lath-shaped magnesium-rich clay mineral with a structure similar to attapulgite. Nagy and Bradley (1955) and Brauner and Preisinger (1956) have proposed structures that differ only in detail. The oxygens which compose the base of the silica tetrahedra form continuous sheets approximately 6.5 Å apart; however, the apex oxygens alternately point above and below the continuous sheet. The tetrahedra which point in the same direction are joined to form two pyroxene chains linked to form an amphibole chain with an extra silica tetrahedron added at regular intervals on each side (Nagy and Bradley, 1955). Brauner and Preisinger's structure has three pyroxene chains linked to form two continuous amphibole chains.

The sheets formed by the apices of the tetrahedra are completed by hydroxyls and magnesium ions in octahedral coordination link the sheets. The one structural arrangement has nine octahedral sites and the other only eight. Both structures have channels on both sides and top and bottom of each ribbon which contains water molecules (zeolitic water). Additional water is bound to the edge of the ribbons and hydroxyls occur in the structure proper.

The ideal structural formula for sepiolite based on the Nagy-Bradley model is $(Si_{12})(Mg_9)O_{30}(OH)_6(OH_2)_4 6H_2O$ and $(Si_{12})(Mg_8)O_{30}(OH)_4(OH_2)_4 8H_2O$ for the Brauner-Preisinger model. Table LVI shows the chemical compositions and Table LVII the structural formulas of nine sepiolite samples (based on the Brauner-Preisinger model). Most calculated structural formulas for sepiolite indicate a minor amount (0.04 to 1.05) of Al^{3+} and/or Fe^{3+} substituting for Si^{4+} (11.96—10.95) in the tetrahedral sheet. The magnesium-rich sepiolites have a relatively consistant composition. Al_2O_3, Fe_2O_3 and in some samples Mn_2O_3 are commonly present in amounts less than one percent; the MgO content ranges from 21 to 25%. Mg fills 90—100% of occupied octahedral positions. Most of the analyses indicate sufficient cations to fill approximately eight (7.74—8.14) octahedral sites. These would fill all the sites allotted by Brauner and Preisinger and leave one vacant site if the Nagy-Bradley structure is correct. The average octahedral cation charge is 15.99 with a relatively small range of values.

Martin-Vivaldi and Cano-Ruis (1955) found from 50 sepiolite analyses a mean value of 53.9% SiO_2 with a standard error of ± 1.9.

"The mean number of octahedral cations caluculated as RO (RO = moles MgO + moles Fe_2O_3 + Al_2O_3 + FeO + MnO), (the latter expressed as MgO) is 0.6 moles per 100 g, equivalent to about

24% MgO. From Nagy–Bradley's (1955) structural formula, $6SiO_2 \cdot MgO \cdot 4H_2O$, adding four extra molecules of water to make it comparable with the analytical data, we obtain for the silica 54.2% and for the MgO 24.2%."

Their data indicate that the moles of MgO per 100 g range from approximately 0.4 to 0.7 with a modal value near 0.6. The moles of $(Al_2O_3 + Fe_2O + FeO + Mn)$ range from approximately 0.0 to 0.15. These data also indicate that eight octahedral sites are filled and in only a few samples is substitution of Mg by trivalent cations enough to account for the filling of only seven octahedral sites.

TABLE LVI

Chemical compositions of some sepiolite samples

	1	2	3	4	5	6	7	8	9	10
SiO_2	52.50	56.10	54.97	52.00	52.97	52.43	45.82	46.60	50.40	50.80
Al_2O_3	0.60	0.42	0.26	0.40	0.86	7.05		0.65		0.66
TiO_2							0.05	0.05		0.02
Fe_2O_3	2.90	0.20	0.21	0.21	0.70	2.24	21.70	16.76	0.73	1.85
FeO	0.70	0.05				2.40	0.20	1.50		1.51
Mn_2O_3					3.14					
MnO							0.17	0.32		
CaO	0.47	0.34					0.90	0.71		
MgO	21.31	24.30	25.35	23.35	22.50	15.08	12.32	15.49	20.28	16.01
NiO									9.78	
CuO					0.87					
Na_2O		1.13	0.09							8.16
NH_3						0.58				
H_2O^-	12.06	10.00	9.25		8.80	10.48	9.48	8.12	9.92	13.68
H_2O^+	9.21	9.21	10.04		9.90	9.45	9.41	10.30	8.63	6.82
Total	99.75	99.*	100.25**		99.74	99.71	100.05	100.50	99.79	99.51

1. Caillère (1936b): sepiolite, Ampandrandava, Madagascar.
2. Hathaway and Sachs (1965): sepiolite, Mid-Atlantic Ridge; analysts Paul Elmore, Sam Botts, Gillison Chloe, Lowell Artis and H. Smith.
3. Maksimovic and Radukic (1961): sepiolite, Goles, south Serbia.
4. Abdul-Latif and Weaver (1969): sepiolite, Asia Minor, Turkey.
5. Nagy and Bradley (1955): sepiolite, Little Cottonwood, Utah, U.S.A.
6. Rogers et al. (1956): aluminous sepiolite, Tintinara, South Australia.
7. Caillère (1936b): xylotile, Sterzing, Tyrol.
8. Caillère (1936a): xylotile, Schneeberg.
9. Caillère (1936a): nickeliferous sepiolite, Nouvelle Caledonie.
10. Fahey et al. (1960): loughlinite, Sweetwater Co., Wyo., U.S.A.

*includes 0.11% K_2O, 0.12% P_2O_5, 0.27% CO_2; **includes 0.02% K_2O, 0.76% CO_2.

Although the Mg-rich variety is the most common, sepiolite-type minerals afford a wide range of compositions which is mostly accommodated in the octahedral sheet. An aluminous sepiolite described by Rogers et al.(1956) has 19% of the octahedral positions filled with Al and 26% of the charge is due to Al. Xylotile, an iron-rich sepiolite, can have up to 9% of the tetrahedral sites and 39% of the filled octahedral sites filled with Fe3+. Fe accounts for 49% of the charge. A nickeliferous sepiolite has been described by Caillère (1936b) containing 9.78% NiO_2 which was assigned to the octahedral sheet (Table LVII). Fahey et al. (1960) have described a variety of sepiolite which contains 8.16% Na_2O and only 16.01% MgO (Table LVI). Leaching experiments indicate sodium substitutes for magnesium: 2 Na for 1 Mg. Assuming all the Na is in the octahedral sheet, a value of 9.79 is obtained for octahedral cations. Preisinger (1959) has suggested that half the Na ions substitute for the Mg ions at the edge of the octahedral sheet, the other half occurs in the channels surrounded by $6 H_2O$ molecules.

Most of the cations in the octahedral positions are the large variety; Al, a smaller cation, is relatively uncommon — this in contrast to palygorskite, where Al commonly

TABLE LVII

Sepiolite structural formulas

	1	2	3	4	5	6	7	8	9	10
Octahedral										
Al		0.06		0.01		1.37				
Fe^{3+}	0.47	0.03		0.04	0.02	0.37	2.69	1.80		0.32
Fe^{2+}	0.13					0.44	0.04	0.29		0.29
Mg	7.14	7.22	8.14	7.96	7.42	4.90	4.20	5.33	6.97	5.53
Mn					0.53		0.03	0.06		
Ni									1.03	
Na										3.65
	7.74	7.81	8.14	8.01	7.97	7.18	6.96	7.48	8.00	9.79
Oct. charge	15.95	15.71	16.28	16.07	15.96	15.90	16.61	15.76	16.00	16.25
Tetrahedral										
Al	0.16	0.04	0.06	0.10	0.24	0.48		0.19		
Fe^{3+}	0.04		0.04		0.09		1.05	1.10	0.15	
Si	11.80	11.96	11.95	11.90	11.67	11.52	10.95	10.71	11.85	11.79
Interlayer										
Ca	0.11	0.15					0.22	0.06		
Cu					0.15					
NH_4						0.43				
Mg									0.15	

For legend, see Table XLVI

TABLE LVIII

Observed water losses (%) compared with those expected from ideal formulas (After Caillère and Hénin, 1961b)

	1	2	3	4	5
Zeolitic water, $< 250°C$	11.43	13.6	11.7	8.1	11.1
Bound water, $250-620°C$	5.7	5.6	5.2	5.4	5.5
Hydroxyls, $620-1000°C$	2.4	2.4	2.6	4.1	2.7
Total	19.53	71.6	19.5	17.6	19.3

1. Ampandrandava, Madagascar.
2. Eski Chehir, Asia Minor.
3. Salinelles, France.
4. Ideal sepiolite (Nagy and Bradley, 1955).
5. Ideal sepiolite (Brauner and Preisinger, 1956).

fills half the octahedral positions. Table LVIII, from Caillère and Hénin (1961b) compares the observed amount of water loss for a number of sepiolites with calculated values. The comparison is much better than for attapulgite. It appears that most of the exchange capacity is due to the replacement of silicon by trivalent ions, though this is not certain. The reported exchange capacity ranges between 20 and 45 mequiv./100 g.

Sepiolite is commonly associated with attapulgite, montmorillonite, dolomite and magnesite. It is reported to have formed in lacustrine environments of high basicity (Longchambon and Morgues, 1927; Millot, 1949), pluvial lakes (Parry and Reeves, 1968), in highly saline evaporitic environments (Yarzhemskii, 1949), under basic marine conditions (Millot, 1964), by hydrothermal alteration of serpentine (Midgley, 1959) by hydrothermal alteration in a deep marine environment (Hathaway and Sachs, 1965) and by the solution of calcite and phlogopite (Lacroix, 1941). Sedimentary sepiolite is most commonly formed along with dolomite in highly alkaline evaporatic environments. Sepiolite apparently forms under more alkaline conditions than does attapulgite and where Si and Mg concentrations are high and Al low.

Chapter 11

KAOLINITE

The two layer silicates are divided into the kaolinite (dioctahedral) and serpentine (trioctahedral) subgroups. The dioctahedral minerals are hydrous aluminum silicates containing minor amounts of other constituents. The trioctahedral minerals vary widely in composition and isomorphous substitution is common; however, these minerals are relatively rare and chemical data are limited.

Kaolinite is by far the most abundant species of the kaolinite subgroup. Although hundreds of chemical analyses of this clay have been made, there is still little known for certain about the exact composition of most (all?) samples. The ideal composition for the kaolinites $Al_4(Si_4O_{10})(OH)_8$ is: 46.54% SiO_2, 39.5% Al_2O_3, 13.96% H_2O; however, in nature, this exact composition is seldom, if ever, found.

Fe_2O_3, TiO_2, MgO, and CaO are nearly always present in kaolinite samples and K_2O and Na_2O are usually present. Most samples either have excess SiO_2 or Al_2O_3. Mineral impurities such as quartz, anatase, rutile, pyrite, limonite, feldspar, mica, montmorillonite, and various iron and titanium oxides are commonly present in addition to a number of other minerals. Si and Al, in the form of hydroxides, apparently can occur as coatings on the kaolinite layers. Although many of these impurities are usually identified, seldom is the analysis sufficiently quantitative to determine if all the deviation from the ideal composition is due to these impurities.

Ross and Kerr (1931) reported on a series of kaolinites in which the SiO_2 / R_2O_3 mole ratio varies from 294:100 to 185:100 (Table LIX); the theoretical value is 200:100. The samples with a ratio near 3:1 are called anauxite. These clays give an X-ray powder pattern similar to that of kaolinite. Hendricks (1942) suggested that neutral silica layers consisting of double sheets of SiO_4 tetrahedra joined by their apices are interstratified irregularly among normal kaolinite layers.

More recently Langston and Pask (1969) demonstrated that the excess SiO_2 was readily soluble and was present as amorphous silica bonding kaolinite flakes together, making them appear as single crystals. Single crystal X-ray diffraction of these crystals indicate they are identical to kaolinite (Bailey and Langston, 1969). The amorphous silica content of these samples ranged from 0.2 to 27.5 percent. The term *anauxite* has no validity as a mineral name.

Alumina is commonly present in excess in kaolinites, usually in amounts of 1—2%. Although some of this may be substituting for Si in the tetrahedral sheet, it is likely that much of it is adsorbed on the surface and edge of the kaolinite flakes as Al

TABLE LIX

Chemical analyses of kaolinite and anauxite (After Ross and Kerr, 1931)

	1	2	3	4	5	6	7	8	9	10	11	12	13	14
SiO_2	54.32	53.80	52.46	48.80	45.56	44.81	45.44	44.70	44.74	43.64	44.92	44.06	44.26	43.78
Al_2O_3	29.96	32.48	32.20	35.18	37.65	37.82	38.52	38.64	37.97	38.33	40.22	39.44	40.22	40.06
Fe_2O_3	2.00	1.12	1.69	1.24	1.35	0.92	0.80	0.96	1.44	1.43	0.54	0.80	0.30	0.64
MnO	—	—	—	—	none	none	—	none	none	—	—	trace	none	—
MgO	0.14	0.26	none	none	0.07	0.35	0.08	0.08	0.06	1.02	0.14	0.26	0.18	0.16
CaO	0.32	0.34	0.03	0.22	0.10	0.43	0.08	0.24	0.09	1.48	0.08	0.06	0.32	0.36
K_2O	none	n.d.	0.31	0.40	0.11	n.d.	0.14	0.14	0.16	n.d.	n.d.	n.d.	n.d.	n.d.
Na_2O	0.37	n.d.	0.25	0.25	1.16	n.d.	0.66	0.62	0.76	n.d.	n.d.	n.d.	n.d.	n.d.
TiO_2	—	—	0.55	0.61	0.19	0.37	0.16	0.22	0.27	—	—	—	none	—
H_2O^-	0.84	0.94	1.38	1.16	0.76	1.10	0.60	0.64	0.58	0.60	0.08	1.06	0.64	1.02
H_2O^+	11.80	10.98	12.07	12.81	13.66	14.27	13.60	13.88	13.98	13.64	14.22	14.16	14.16	14.08
Total	99.75	99.92	100.94	100.67	100.61	100.07	100.08	100.12	100.05	100.14	100.20	99.84	100.08	100.10
SiO_2/R_2O_3	294:100	274:100	267:100	230:100	202:100	199:100	197:100	195:100	195:100	189:100	189:100	188:100	186:100	185:100

n.d. = not determined

1. Anauxite, Bilin, Czechoslovakia; analyst William F. Foshag.
2. Anauxite, Bilin, Czechoslovakia; analyst F.A. Gonyer.
3. Anauxite, Mokelumne River, 1 mile west of Lancha Plana, Calif., U.S.A.; analyst J.G. Fairchild.
4. Anauxite, Newman Pit, near Ione, Calif., U.S.A.; analyst J.G. Fairchild.
5. Kaolinite from Sand Hill Station, near Pontiac, S.C., U.S.A.; analyst F.A. Gonyer.
6. Kaolinite associated with arkosic sand from Mexia, Texas, U.S.A.; analyst F.A. Gonyer.
7. Kaolinite in vermicular grains from a weathered pegmatite, Roseland, Va.; U.S.A.; analyst F.A. Gonyer.
8, 9. Kaolinite from locality 1 mile south of Ione, Amador County, Calif., U.S.A.; analyst F.A. Gonyer.
10. Kaolinite, Abatik River, northern Alaska; analyst F.A. Gonyer.
11. Kaolinite, Jerome, Ariz., U.S.A.; analyst F.A. Gonyer.
12, 14. Kaolinite, in vermicular grains from kaolin seams in bauxite, Saline County, Ark., U.S.A.; analyst F.A. Gonyer.
13. Kaolinite from Franklin, N.C., U.S.A., from a typical kaolin derived from weathered pegmatite; analyst F.A. Gonyer.

hydroxide, perhaps, in part, acting as a cement. Thomas and Swoboda (1963) found that hydroxy-Al ions were present equivalent to 10—15% of the total C.E.C.

Chemical dissolution techniques indicate kaolinite from Cornwall contains 3.1—4.9% of easily soluble SiO_2 and 1.5—5.9% of easily soluble Al_2O_3 (Follett et al., 1965). Most of this material is presumably present as amorphous material. Experiments (by the senior author) with Georgia kaolinite indicate the amount of amorphous material varies as a function of particle size and preparation (Table LX). Amorphous silica and alumina is a common constituent of kaolinite and considerable care must be taken in determining and interpreting the significance of the Si/Al ratio of kaolinites.

TABLE LX

Material soluble in 0.5 N NaOH after boiling 2.5 min.

	% SiO_2	% Al_2O_3
Crude	2.1	0.95
Coarse books	1.28	0.57
Commercial fine fraction	3.8	1.5
Moderate physical treatment	13.4	5.3

A number of kaolinites were analyzed by Van der Marel (1958) and such parameters as crystallinity, surface area, and cation exchange capacity determined (Table LXI). Though most of the measured parameters show good negative or positive correlation with each other, they show little correlation with composition. As would be expected, H_2O^- increases as surface area increases (Fig.21) and can be used to give a reasonable determination of surface area. The limited data show an increase in TiO_2 as surface area increases up to a maximum surface area of 35—40 m^2/g (Fig.22). This suggests that much of the TiO_2 is adsorbed. The range of SiO_2/Al_2O_3 ratios are similar to those reported by Ross and Kerr (1931). The total "chemical impurities" range from 5 to 15%. How much of this impurity is in the kaolinite layer was not established. Van der Marel believes the decrease in intensity of recorded data (X-ray diffraction, D.T.A., infrared) with increase in surface area is due to the presence of amorphous layers.

A number of chemical analyses of Georgia kaolinites with varying particle size were supplied by Dr. H.H. Murray of the Georgia Kaolin Company. The amount of less than 1μ material in the samples ranged from 11% to 96%. Al_2O_3 tends to be relatively more abundant in the coarser clays. Both Fe_2O_3 and TiO_2 tend to increase (0.26 to 1.10 and 1.03 to 4.00, respectively) as particle size decreases although the relation is

TABLE LXI

Chemical analyses of kaolinites

	1	2	3	4	5	6	7	8	9	10	11	12	13	14	15	16	17
SiO_2	44.15	42.42	43.60	41.57	45.81	41.03	42.81	44.62	43.84	44.35	41.56	42.10	41.76	43.8	44.87	46.86	45.1
Al_2O_3	38.99	38.74	37.91	38.87	33.90	38.40	37.67	35.43	38.85	32.32	36.22	33.17	37.86	39.1	36.32	38.22	37.8
H_2O^+	12.96	13.68	13.62	13.53	12.62	14.05	13.29	12.84	13.54	11.63	12.83	12.45	13.28	15.25	13.60	13.59	13.4
Total	96.10	94.84	95.13	93.97	92.33	93.48	93.77	92.89	96.23	88.30	90.61	87.72	92.90	98.15	94.79	98.67	96.3
Fe_2O_3	1.06	0.69	0.25	0.60	3.57	0.59	0.86	0.93	0.46	4.26	1.68	5.68	0.99	0.90	0.35	1.19	0.45
TiO_2	0.62	0.88	1.96	1.26	1.36	1.60	2.32	0.78	1.03	2.48	1.64	1.46	2.01	–	1.04	0.41	2.1
MgO	0.14	0.46	0.05	0.63	0.02	0.22	0.20	0.45	0.36	0.32	0.36	0.46	0.28	–	0.14	–	0.09
Others	1.59	2.60	1.89	2.99	2.11	2.46	1.83	3.03	0.77	3.07	3.06	2.73	2.53	0.58	0.41	–	1.01
SiO_2/Al_2O_3	1.93	1.86	1.95	1.82	2.30	1.82	1.93	2.14	1.92	2.33	1.95	2.16	1.88	1.90	2.10	2.11	2.03
$SiO_2/(Al_2O_3 + Fe_2O_3)$	1.89	1.84	1.94	1.80	2.15	1.80	1.91	2.11	1.90	2.15	1.89	1.94	1.85	1.82	2.09	2.07	2.01

*1. Cornwall, England.
 2. Mesa Alta, U.S.A.
 3. Bath, S.C., U.S.A.
 4. Bangka, Indonesia.
 5. Gabon, Belgian Congo.
 6. Murfreesboro, U.S.A.
 7. Dry Branch, U.S.A.
 8. Zettlitz, Czechoslovakia.
 9. Macon, U.S.A.
10. Dhong Tuan, Thailand.
11. Reims, Provence, France.
12. Brokoponda, Surinam.
*13. New Jersey, U.S.A. (fire clay).
14. New Jersey, U.S.A. 100–200 mesh books: Lodding (1965).
15. Jamaica; Worrall and Cooper (1966).
16. South Africa (flint clay): Warde (1950).
17. Kentucky, U.S.A. (flint clay): Patterson and Hosterman (1962).

*1–13 from Van der Marel (1958).

KAOLINITE 135

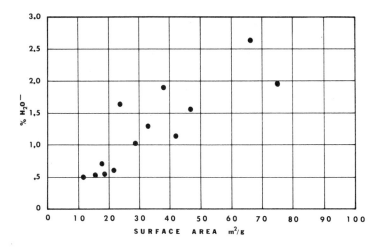

Fig. 21. The relation of surface area to percent H_2O of thirteen kaolinites. (Data from Van Der Marel, 1958.)

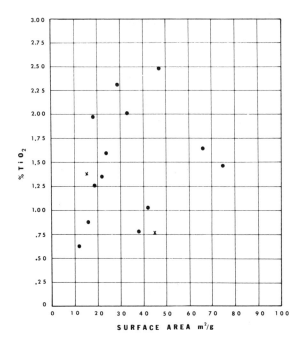

Fig. 22. The relation of surface area to percent TiO_2 of fifteen kaolinites. Two distinct trends are suggested.

TABLE LXII

TiO$_2$ content of fractionated Georgia kaolinite

Size of fraction (in μ)	% TiO$_2$
10$^+$	0.08
5–10	1.0
2.5–5	1.2
1–2.5	1.7
0.5–1	1.9
0.5	2.6

poor.

Smith (1929) reported that 115 kaolinite samples from Georgia had an average of 1.13% TiO$_2$. The values ranged from 0.5 to 2,2% TiO$_2$.

Nagelschmidt et al.(1949) determined that the TiO$_2$ in Georgia kaolinite was present as anatase and showed that it increased in abundance as the particle size decreased (Table LXII).

The senior author's study of kaolinite from east Georgia showed an opposite trend for TiO$_2$ in the material finer than $2\,\mu$ (Table LXIII). The material coarser than $2\,\mu$ has less TiO$_2$ than the $1-2\,\mu$ fraction.

Thus, although there is a tendency for finer grained kaolinite deposits to have more TiO$_2$ than coarser grained deposits, the association does not necessarily apply to individual samples. Most of the TiO$_2$, in the form of anatase, can be dispersed and flocculated independent of the kaolinite and the reported distributions of TiO$_2$ are a function of laboratory treatment more than anything else. The anatase contains signi-

TABLE LXIII

Chemical data on east Georgia kaolinite

Size of fraction (in μ)	TiO$_2$	Fe$_2$O$_3$	MgO
1–2	1.68	1.08	0.87
0.5–1.0	1.37	1.07	0.77
0.25–0.5	1.33	1.04	0.80
0.10–0.25	1.26	1.05	0.80
<0.01	1.10	1.10	0.77

ficant amounts of Fe and Mg although most of the Fe and Mg is present in other forms.

Konta (1969) described kaolinites from West Bohemia in which the TiO_2 content is as low as 0.02%. The kaolinite is apparently a residual weathering product from a leucocratic granite, high in muscovite. The TiO_2 content of kaolinite is apparently dependent on the source mineral. When feldspar is the major source or the clay is of hydrothermal origin, the TiO_2 content of the kaolinite is low; when the source is a biotite schist or biotite granite the TiO_2 content is relatively high.

Magnetic and flotation techniques can reduce the TiO_2 content of Georgia kaolinites from 1.7–2.0% to 0.2–0.4%. Leaching experiments by Dolcater et al. (1970) indicate that on an average 15% (range 0–30%) of the Ti in seven samples (six from Georgia) was in the kaolinite structure. Thus, present data indicate that some Ti does occur in the kaolinite layer, presumably in the octahedral sheet.

Smith (1929) found that the ferric oxide content of 115 Georgia kaolinites commonly ranged from 1 to 2% and averaged 1.43%. Many of the Georgia kaolinites have Fe_2O_3 values in the 0.1–3.0% range but relatively few have more than 2%. Much of the iron is present as iron oxides and pyrite; however, some is complexed with the TiO_2 and some occurs in the micas and montmorillonites commonly present. It has not been established that any of the iron is in the kaolinite structure; however, after removing most of the mineral impurities and leaching, the clays still retain about 0.3% Fe_2O_3 suggesting that this amount is probably present in the structure. Mössbauer analyses and leaching studies (Malden and Meads, 1967) showed that 0.3% Fe_2O_3 was present in the structure of the kaolinites from St. Austill; an additional 0.2–0.4% is leachable and presumably present as Fe oxides and hydroxides, and an indeterminant amount was present in mica (removed by magnetic techniques).

Most kaolinites contain appreciable amounts of MgO (range 0.01–1.0%; modal value between 0.2–0.3%). Bundy et al. (1965) found that MgO as well as total iron and soluble iron were directly related to the C.E.C. and suggested that the Mg and Fe were present in montmorillonite which they believe is commonly present, in amounts less than 5%, in kaolinites. This may be true in part, but electron probe studies (Weaver, 1968) indicate that some of the MgO is related to the TiO_2-Fe_2O_3 material and some is present in biotite.

Robertson et al. (1954) analyzed two kaolinites in detail and concluded that Fe was present in the octahedral sheet and that there was sufficient isomorphous substitution to account for the cation exchange capacity (Table LXIV). These clays were not pure and it was necessary to make a number of assumptions in order to obtain these results.

In an effort to establish basic differences between Georgia soft kaolinites (clay feels soft and friable) and Georgia hard kaolinites (breaks with a conchoidal fracture and crumbles with difficulty between thumb and finger) Hinckley (1961) analyzed over three hundred samples. Although this is an empirical division it effectively divides

TABLE LXIV

Ionic constitution of two kaolinites (After Robertson et al., 1954)

	Theoretical for kaolinite	Pugu D kaolinite		Pugu K kaolinite	
Mg/2	–	0.002 ⎫		0.002 ⎫	
Ca/2	–	0.004 ⎬ 0.021		0.003 ⎬ 0.009	
Na	–	0.008 ⎪		0.004 ⎭	
H	–	0.007 ⎭		–	
Al	2.0	1.958 ⎫		1.957 ⎫	
Fe^{3+}	–	0.039 ⎬ 2.000		0.037 ⎬ 1.999	
Mg	–	0.003 ⎭		0.005 ⎭	
Al	–	0.018 ⎫ 2.000		0.009 ⎫ 2.008	
Si	2.0	1.982 ⎭		1.991 ⎭	
O	5.0	5.000 ⎫ 9.122		5.000 ⎫ 9.092	
OH	4.0	4.122 ⎭		4.092 ⎭	
C.E.C. in mequiv./100 g		measured 7.4 calculated 7.53		4.6	

the clays on the basis of crystallinity. The soft kaolinites are relatively well-ordered whereas the hard kaolinites are disordered by random shifts along the b-axis. Table LXV is a summary of the results obtained by Hinckley (grit-free samples were used for chemical analyses). He found Al_2O_3 was significantly more abundant in the well-crystallized kaolinite (soft) and Fe_2O_3 in the disordered kaolinites (hard). The SiO_2/Al_2O_3 ratios of both types (hard 194:100; soft 188:100) are below the theoretical value of 200:100. This indicates a deficiency of SiO_2 relative to Al_2O_3 of about 1% in the disordered kaolinites and about 2.5% in well-crystallized kaolinites.

Table LXVI shows the correlation coefficients obtained for the hard and soft types. The existence of two clay populations limits the meaning of correlations found for the combined data (Hinckley, 1961). In both groups K_2O and mica and K_2O and Fe_2O_3 are positively correlated. In addition, for the soft type there is a positive correlation between Fe_2O_3 and mica and negative correlations between mica and books, and Fe_2O_3 and books. These interrelations suggest, but do not prove, that books are derived from the mica and that much of the K_2O and Fe_2O_3 may be present in the mica or that a leaching process that altered the mica and removed its interlayer K_2O also removed the iron regardless of where it occurred (pyrite, anatase, iron oxides, etc.).

TABLE LXV

Differences between hard and soft kaolinites of Georgia (After Hinckley, 1961)

		Mean	Standard deviation	Range	N
SiO$_2$	hard	44.1%	0.49	43.2–45.5	50
	soft	43.8%	0.75	41.1–47.9	197
Al$_2$O$_3$	hard	38.6%	1.07	35.4–40.7	50
	soft	39.7%	1.26	31.9–42.3	197
K$_2$O	hard	0.28%	0.19	0.02–0.76	50
	soft	0.14%	0.14	0.00–1.12	197
Fe$_2$O$_3$	hard	1.9%	0.71	0.75–5.52	50
	soft	0.2%	0.23	0.00–1.64	197
TiO$_2$	hard	1.6%	0.27	1.09–2.12	50
	soft	1.5%	0.48	0.43–3.87	197
Montmorillonite	hard	2.3%	0.78	1.00–3.46	50
	soft	3.0%	6.86	0.01–50.30	197
Crystallinity	hard	0.46	0.13	0.28–0.75	50
	soft	0.90	0.22	0.55–1.43	197
> 200 mesh	hard	2.13%	1.71	0.05–6.54	50
	soft	0.67%	0.69	0.04–2.73	197
Quartz	hard	69.0[a]	65.0	2–189	50
	soft	4.0[a]	9.0	0–64	197
Mica	hard	38.0[a]	58.0	0–171	50
	soft	94.0[a]	60.0	0–85	197
Books	hard	2.0[a]	3.0	0–12	50
	soft	75.0[a]	50.0	13–190	197
Undispersed clay	hard	64.11[a]	75.82	0–196	36
	soft	14.23[a]	20.82	0–100	72
Boron	hard	1.06[b]	0.099	0.91–1.16	8
	soft	0.65[b]	0.264	0.40–0.96	8
Gallium	hard	1.62[b]	0.099	1.47–1.75	8
	soft	1.68[b]	0.063	1.61–1.78	8
MnO	hard	359[c]	150	174–697	12
	soft	356[c]	98	251–613	12
Bulk density	hard	1.620	0.024	1.514–1.734	12
	soft	1.480	0.016	1.454–1.505	12
Exchangeable calcium	hard	72.5[d]	2.60	70–75	12
	soft	120.8[d]	45.81	65–190	12
Exchangeable potassium	hard	31.8[d]	2.30	29–36	12
	soft	33.0[d]	2.41	29–36	12

[a]grains per 200 points; [b]10/% transmission (emission spectroscopy); [c]counts per 40 seconds (X-ray fluorescence); [d]parts per million.

TABLE LXVI

Correlation coefficients of some parameters of hard and soft Georgia kaolinites (After Hinckley, 1961)

	SiO_2	Al_2O_3	K_2O	Fe_2O_3	TiO_2	Montmorillonite	Crystallinity index	Quartz	Mica	Kaolinite books
Simple correlation coefficients: hard grit-free samples										
SiO_2	1.0000	−0.1072	0.4330	−0.1427	0.3180	−0.3228	0.3687	−0.1567	0.4181	0.1388
Al_2O_3		1.0000	−0.5229	−0.4725	0.0755	−0.0180	−0.2123	0.0671	−0.3692	−0.3056
K_2O			1.0000	0.4424	0.0460	−0.1144	0.4489	−0.2094	0.8033	0.4814
Fe_2O_3				1.0000	−0.4791	0.3023	−0.0802	0.0136	0.3503	0.4322
TiO_2					1.0000	−0.8209	0.5695	−0.4883	0.1529	0.0855
Montmorillonite						1.0000	−0.5633	0.5932	−0.3763	−0.3921
Crystallinity index							1.0000	−0.2722	0.3977	0.3339
Quartz		N = 50						1.0000	−0.5006	−0.3763
Mica		$P_{0.05}$ = 0.279							1.0000	0.7273
Kaolinite books		$P_{0.01}$ = 0.361								1.0000
$\bar{X}$	44.0780	38.5600	0.2824	1.8824	1.5006	2.3584	0.4570	69.1600	38.1600	1.8000
s	0.4916	1.0759	0.1935	0.7124	0.2696	0.7764	0.1285	64.8674	58.1860	3.1875
Simple correlation coefficients: soft grit-free samples										
SiO_2	1.0000	−0.4709	−0.0192	−0.0910	−0.2597	0.5129	−0.0103	0.0187	−0.1203	−0.0762
Al_2O_3		1.0000	−0.0712	−0.0604	−0.1868	−0.6205	0.2100	0.0527	0.0469	0.1229
K_2O			1.0000	0.6141	0.0857	−0.0692	−0.2244	0.0795	0.2611	−0.3092
Fe_2O_3				1.0000	−0.0997	−0.0554	−0.3311	0.0152	0.5063	−0.4471
TiO_2					1.0000	−0.1540	0.1497	−0.0796	−0.1499	0.2390
Montmorillonite						1.0000	−0.3248	−0.0745	−0.0882	−0.2336
Crystallinity index							1.0000	0.1909	−0.5536	0.4735
Quartz		N = 197						1.0000	−0.0798	−0.1148
Mica		$P_{0.05}$ = 0.140							1.0000	−0.7894
Kaolinite books		$P_{0.01}$ = 0.184								1.0000
$\bar{X}$	43.8472	39.6975	0.1390	0.2383	1.5046	3.0207	0.9040	3.5280	94.4772	75.4213
s	0.7512	1.2672	0.1372	0.2302	0.4780	6.8600	0.2166	9.1809	60.4349	50.4560

In the hard kaolinites there is a positive correlation between mica and books which Hinckley suggests may be due, in part, to sedimentation conditions which favored the simultaneous deposition of books and mica in preference to quartz. This type also has TiO_2 positively related to crystallinity. Crystallinity increases as grain size increases. This may reflect a leaching process in which quartz, Fe_2O_3 and montmorillonite are removed, TiO_2 is concentrated as anatase, and some authigenic kaolinite is formed.

Hinckley believes the differences between the two types of clays is largely a function of the environment of deposition and post-depositional leaching. The poorly crystallized hard clays were believed to have been deposited in a marine environment where face to face flocculation occurred. The well-crystallized soft clays were presumably flocculated edge to face in a fresh-water environment. The more porous soft clays were more thoroughly leached and re-crystallized than the less porous hard clays. Like other kaolinite studies this one does not provide any specific information about the actual chemical composition of the kaolinite minerals.

Worral and Cooper (1966) analyzed a pure, poorly crystallized kaolinite from Jamaica (Table LXVII). The cation exchange capacity is 24.4 mequiv./100 g. They suggest that substitution in the octahedral sheet is the cause of the high cation exchange capacity and may be the cause of the disorder.

Most kaolinite is formed by the acid leaching of alkaline rocks, primarily the feldspars and micas; however, practically any silicate rock or mineral will alter to kaolinite if leaching conditions are suitable for a sufficiently long period of time. Kaolinites that are formed by weathering and remain in place are called residual kaolinites. Those that are transported and sedimented are called sedimentary kaolinites

TABLE LXVII

Ionic composition of Jamaican kaolinite

Position	Cation	Atoms
Octahedral	Al	1.906
	Fe	0.013
	Mg	0.010
	Ti	0.036
Exchangeable	NH_4	0.065
Tetrahedral	Si	2.002
	O	5.000
	OH	4.000

(additional kaolinite can be formed after deposition). Kaolinite can also form from solution by resilication of aluminum-rich materials and by hydrothermal alteration.

In the samples analyzed by Ross and Kerr (1931) the residual kaolinites have an average Fe_2O_3 content of 0.68% and the sedimentary kaolinites an average of 1.36%. However, the well-crystallized sedimentary kaolinites of Georgia contain an average of only 0.2% Fe_2O_3 and the poorly crystallized ones 1.9%. The famous kaolinite deposits in southwest England:" ... are the result of the attack on the granite by emanations derived from the igneous bodies themselves as a phase of post-consolidation history of the intrusions" (Holmes, 1950). Although these kaolinites are considered to be relatively iron-free, an average of four analyses gives 0.80% Fe_2O_3 and 0.12% FeO. The average TiO_2 value for these samples is 0.21% which is an appreciably lower value than is reported for most kaolinites formed by surface weathering. Most of the Fe_2O_3, TiO_2, CaO, K_2O, MgO, etc., are present as impurities and their relative abundance depends largely upon the composition of the source material, the efficiency of the leaching process, and the chemical conditions in the environments of deposition and post-depositional environments. Thus, general relations would be complex and only in restricted situations is it likely that a direct relation between chemical composition and origin can be established.

Fire clays, ball clays, flint clays are kaolinite-rich clays, usually of the b-axis disordered variety, which contain a relatively high impurity content. Illite, montmorillonite, diaspore, boehmite, quartz, and organic material are the minerals usually associated with these deposits. Few, if any, of the kaolinite minerals in these clays have been concentrated enough to afford meaningful chemical data.

One of the major differences in the reported chemical composition of the kaolinite minerals is in the H_2O^+ and H_2O^- values. In part, these variations may be real but many must be due to the presence of halloysite and other impurities, variation in grain size and surface area, and in the methods of dehydration. H_2O^- increases linearly with increase in surface area and with decrease in grain size.

Kaolinite seldom occurs interstratified with 2:1 clay minerals although Sudo and Hayashi (1956) described a randomly interstratified kaolinite-montmorillonite in acid clay deposits in Japan. On the basis of the chemical analysis (SiO_2 = 41.94, Al_2O_3 = 30.12, Fe_2O_3 = 2.42, FeO = 0.21, MgO = 1.52, CaO = 0.32, TiO_2 = 0.40, H_2O^+ = 11.10, H_2O^- = 12.88, total = 100.91) they concluded the clay was 80% montmorillonite and 20% kaolinite. Although considerable montmorillonite is weathered to kaolinite, this type of interstratification is relatively rare. Schultz et al. (1969) found samples from the Yucatan Peninsula that contained 40–60% kaolinite interlayered with montmorillonite:

$$K_{0.25}Na_{0.20}Ca_{0.20}Mg_{0.15}$$
$$(Al_{1.75}Fe^{3+}_{0.50}Mg_{0.25})(Si_{3.62}Al_{0.38})O_{10}(OH)_2$$

The cation exchange capacity of the kaolinite minerals is relatively low but due to

KAOLINITE

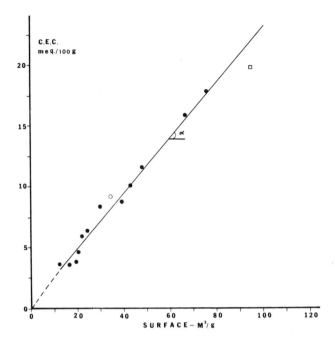

Fig.23. Cation exchange capacity (C.E.C.) in mequiv./100 g of various kaolinite samples compared with their surface in m²/g. (After Van Der Marel, 1958.)

the omnipresent impurities it is difficult to determine true values. Van der Marel (1958) lists values ranging from 3.6 to 18.0 mequiv./100 g which is the generally accepted range. He also showed (Fig.23) an excellent linear relation between C.E.C. and surface area (other analyses fall on this same trend). He calculated that there was one monovalent cation per 67 A^2, which is close to the values reported for clays of much higher exchange capacity (illite and montmorillonite = 75 to 100 A^2). Van der Marel suggested that an amorphous layer might be present on the kaolinite particles and account for the close relation between surface area and C.E.C.

Part of the exchange capacity is due to broken bonds at the edges of the flakes. This charge is reversible, being negative in a basic environment and positive in an acid environment and zero at neutrality (Schofield and Samson, 1953). Van der Marel demonstrated that the exchange capacity at pH 8.2 was 19% larger than that at neutrality and assumed this increase was due to edge charge. Sumner (1963) has noted that iron oxides (and aluminum hydroxides) are present on the surface of kaolinite and materially influence the measured C.E.C. Iron oxides behave amphoterically. Thus, in a basic solution iron oxide acts to increase the C.E.C. of a kaolinite-iron oxide complex and in an acid solution tends to produce a low value. He demonstrated that the negative charge on kaolinite was constant between pH 2.4 and 5.0, but above this increased with increasing pH. He concluded, along with Schofield (1949) and others, that this indicated the kaolinite has a permanent negative charge below pH 5 due to isomorphous replacement within the structure.

In a study of Georgia kaolinites Bundy et al. (1965) stated that they believed much of the exchange capacity of kaolinite to be due to the presence of minor amounts of montmorillonite (identified by Mg content). Approximately 0.8 mequiv./100 g was the highest exchange capacity measured for kaolinites which they believed contained no montmorillonite. This could be accounted for entirely by edge charge.

It is assumed that some exchange is due to isomorphous substitution but this has not been proven. Schofield and Samson (1953) calculated that only one Al^{3+} need replace one Si^{4+} in 400 unit cells to afford an exchange capacity of 2 mequiv./100 g. There is enough excess Al^{3+} in most kaolinites to account for 10 times this exchange capacity. Thus it appears likely that most of the excess Al^{3+} does not substitute in the tetrahedral sheet. The iron-rich kaolinite described by Kunze and Bradley (1964) has an exchange capacity of 60 mequiv./100 g; however, it is likely that much of this is due to the presence of iron oxides.

The identity of the exchangable cations is seldom determined although Hinckley (1961) showed that in the Georgia kaolinites exchangable Ca was two to five times as abundant as K. Na and Mg were not determined. Hydrogen is presumably the most abundant exchangable cation at the time of formation.

Chapter 12

DICKITE AND NACRITE

The minerals dickite and nacrite have a theoretical composition identical to kaolinite, but have their layers stacked in alternating regular sequences. Dickite has a 2-layered monoclinic cell (Newnham and Brindley, 1956) and nacrite a 6-layered monoclinic cell (Hendricks, 1939a).

Nacrite, the rarest of the kaolinite minerals, probably forms only by hydrothermal activity (Table LXVIII). Dickite is formed by hydrothermal solutions and occurs quite commonly as a secondary clay in the pores of the sandstones (Smithson and Brown, 1954), also in coal beds (Honess and Williams, 1935) and in geodes (Ross and Kerr, 1931). Although conditions of the origin of dickite vary, there is little significant variation in composition (Table LXIX). SiO_2/Al_2O_3 ratios are near the theoretical value of 200:100 although most samples show a deficiency of SiO_2 as do most kao-

TABLE LXVIII

Chemical analyses of nacrite

	1	2	3
SiO_2	44.75	45.91	46.32
Al_2O_3	39.48	39.65	39.65
Fe_2O_3	0.53	–	0.05
MgO	0.19	–	trace
CaO	0.13	–	0.08
CaF_2	–	0.68	–
TiO_2	–	–	trace
H_2O^+	14.40	13.77	14.21
H_2O^-	0.61	–	none
Total	100.09	100.01	100.31
SiO_2/Al_2O_3	192:100	196:100	199:100

1. Ross and Kerr (1931): hydrothermal, associated with galena crystals, Brand, Saxony; analyst F.A. Gonyer.
2. Ross and Kerr (1931): in granite pegmatite, St. Peters Dome, Colo., U.S.A.
3. Von Knorring et al. (1952): probably hydrothermal in kyanite-bearing quartzite, Hirvivaara, Finland; analyst O. von Knorring.

TABLE LXIX

Chemical analyses of dickite

	1	2	3	4	5	6
SiO_2	45.04	46.35	46.43	45.4	46.14	47.4
Al_2O_3	40.70	39.59	39.54	39.2	39.61	38.0
Fe_2O_3	trace	0.11	0.15	–	–	–
MgO	–	–	0.17	0.3	–	–
CaO	0.22	–	none	trace	–	–
TiO_2	trace	–	none	none	–	0.2
H_2O^+	14.08	13.93	14.20	13.4	13.91	13.7
H_2O^-	none	–	–	0.4	–	0.0
F	–	0.15	–	–	–	–
O/F	–	0.16	–	–	–	–
Total	100.04	100.13	100.51[1]	100.1[2]	99.66	99.3
SiO_2/Al_2O_3	188:100	194:100	199.6:100	197:100	198:100	212:100

[1] includes: 0.02% K_2O, 0.03% Na_2O; [2] includes: 1.0% carbonaceous matter.

1. Ross and Kerr (1931): fine grained, presumably hydrothermal, mining camp in Cusihuiriachic, Chihuahua, Mexico; analyst J.G. Fairchild.
2. Ross and Kerr (1931): hydrothermal, from National Belle mine, Ouray, Colo., U.S.A.; analyst W.F. Hildebrand.
3. Schmidt and Heckroodt (1959): hydrothermal, elongated crystals, Barkly East, Union of S. Africa.
4. Dunham et al. (1948): coating on joints of dolomitic mudstone, probably hydrothermal, magnesium limestone, Durham, England; analyst C.O. Harvey.
5. Honess and Williams (1935): pocket in coal mine, Pine Knot Colliery, Schuykill Company, Pa., U.S.A.; analyst G.A. Brady.
6. Smithson and Brown (1957): pores in sandstone, authigenic, Millstone Grit, Anglesey, England.

linite values. Ross and Kerr (1931) report ratios as low as 183:100 in a fine-grained dickite. These low values suggest that extraneous Al_2O_3 might be present and undetected. The authigenic dickite which forms in sandstones has a relatively high ratio of 212:100. Such an environment might be expected to favor a high silicon value although there are a number of different forms in which it could be present.

In comparing samples of dickite (2) and kaolinite (6), McLaughlin (1959) found that, in general, Ti, Na, K, Li, Zr, Y, and Sr are lower in dickite samples than in either kaolinite- or halloysite-bearing materials. Mg, Pb, and Ba seemed higher in dickite than in the others. Additional data suggest some of these differences are not real and those that are, are due to contaminants such as mica, limonite and zircon. Hydrothermal kaolinite and dickite both appear to contain fewer contaminants than secondary or authigenic kaolinites.

Only a few analyses of nacrite have been made. These have SiO_2/Al_2O_3 ratios near the theoretical value of 2:1 but do not appear to be appreciably more chemically pure than most dickites. Ti is lower in both nacrite and dickite than in most kaolinites, but the difference in Fe content is less distinct. Much of the Fe in nacrite and dickite is apparently secondary and its concentration relative to Ti is due to its high mobility.

Chapter 13

HALLOYSITE

Halloysite has the same theoretical SiO_2/Al_2O_3 ratio as kaolinite, but the water content is higher. The fully hydrated variety, $Al_2O_3 \cdot 2SiO_2 \cdot 4H_2O$, contains two water layers between adjacent kaolinite layers resulting in a 10.1 Å thick unit. Water in excess of about 2.2–2.3 H_2O is lost at temperatures as low as 60°C. To lower the water content to $2H_2O$ and reduce the thickness to the kaolinite value of 7.2 Å it is necessary to heat the clay at temperature of from 200 to 400°C. Brindley (1961b) refers to the three types as hydrated halloysite, metahalloysite and dehydrated halloysite.

Halloysite has a tubular morphology (Bates et al., 1949; 1950) with the *c* axis radial. Most halloysites are highly disordered along both the *a* and *b* axes and around the circumference. More recent studies have indicated that some tubular halloysites have an appreciable degree of structural regularity (Honjo et al., 1954; Takahashi, 1958; De Souza Santos et al., 1965).

Ross and Kerr (1934) listed 13 chemical analyses of halloysite. They stated that "with materials as fine grained as most halloysites it is not possible to apply any treatment to remove impurities. For that reason, nearly all the samples analyzed represent material that occurs naturally in a perfectly homogeneous and apparently pure condition." Bates (1959) lists 18 halloysite analyses (Table LXX) including the 13 from Ross and Kerr's paper.

Halloysite analyses show more variability than analyses of the other kaolinite minerals. Part of this may be real, but part is due to the fine-grained poorly crystalline nature of the material which makes impurities more difficult to detect. The content of low-temperature water depends largely on the dehydration history of the clay and has little meaning otherwise. High-temperature water in all but two samples is larger than the theoretical value of 13.96%. The high value is believed due to water trapped between layers during dehydration. The SiO_2/Al_2O_3 ratio ranges from 165:100 to 206:100 although most values fall in the 180:100 to 200:100 interval. The average ratio for 22 samples is 188:100 which is identical to the ratio reported for the well-crystallized kaolinites of Georgia (Hinckley, 1961).

The SiO_2/Al_2O_3 ratios for the various kaolinite minerals suggest that most of the variations are a function of impurities. Ninety percent of the analyses show a ratio less than 2:1 suggesting that an excess of Al_2O_3 is much more common than an excess of

TABLE LXX

Chemical analyses of halloysites*

	1	2	3	4	5	6	7	8
SiO_2	44.75	40.26	43.67	44.34	44.08	44.08	41.62	44.68
Al_2O_3	36.94	37.95	37.91	37.39	39.20	38.60	38.66	38.59
Fe_2O_3	0.31	0.30	0.26	0.42	0.10	0.32	0.62	0.39
MnO	–	–	–	–	–	–	–	–
MgO	–	–	trace	0.04	0.05	0.22	0.08	0.08
CaO	0.11	0.22	–	0.17	none	0.12	0.10	0.18
K_2O	(	)	–	0.04	( )	–	–	0.05
	(0.60	0.74)			(0.20)			
Na_2O	(	)	–	0.17	( )	–	–	0.11
TiO_2	–	–	–	–	–	–	–	–
H_2O^-	2.53	4.45	3.79	2.00	1.44	2.34	4.26	1.55
H_2O^+	14.89	15.94	14.50	15.09	14.74	14.72	14.64	14.90
FeO	–	–	–	–	none	–	–	–
P_2O_5	–	–	0.12	–	–	–	–	–
Total	100.13	99.86	100.25	99.66	99.81	100.40	99.98	100.53
SiO_2/Al_2O_3	206:100	180:100	195:100	201:100	191:100	189:100	177:100	196:100

1. Ross and Kerr (1934): Liege, Belgium; analyst L.T. Richardson.
2. Ross and Kerr (1934): Huron County, Ind., U.S.A.; analyst L.T. Richardson.
3. Ross and Kerr (1934): Huron County, Ind., U.S.A.; analyst L.T. Richardson.
4. Ross and Kerr (1934): Peppers, N.C., U.S.A.; analyst E.T. Erickson.
5. Ross and Kerr (1934): Hickory, N.C., U.S.A.; analyst J.G. Fairchild.
6. Ross and Kerr (1934); Hickory, N.C., U.S.A.; analyst F.A. Gonyer.
7. Ross and Kerr (1934): Adams County, Ohio, U.S.A.; analyst F.A. Gonyer.
8. Ross and Kerr (1934): Brandon, Rankin County, Miss., U.S.A.; analyst Charles Milton.
9. Ross and Kerr (1934): Leakey, Real County, Texas, U.S.A.; analyst R.K. Bailey.
10. Ross and Kerr (1934): Sneeds Creek, Newton County, Ark., U.S.A.; analyst F.A. Gonyer.
11. Ross and Kerr (1934): Franklin, N.C., U.S.A.; analyst F.A. Gonyer.

SiO_2. Much of this Al_2O_3 is probably present as aluminum hydroxides adsorbed on the surface and between the layers of the kaolinite particles.

The average Fe_2O_3 content of the 19 halloysite samples in Table LXX is 0.43%. This value is lower than the average value for kaolinites but is considerably higher than the value for the well-crystallized kaolinite clays of Georgia. The TiO_2 content is not often determined, but when analyzed, it is usually present in amounts less than 0.1% (average 0.08% for eight analyses). This is only 1/10 to 1/20 the amount found in most residual and sedimentary kaolinites. Halloysites appear to have a lower average TiO_2 content than any other clay mineral species and a much lower content than

9	10	11	12	13	14	15	16	17	18	19
44.50	39.84	44.18	43.10	45.20	44.51	44.35	39.22	43.6	44.3	44.64
38.68	39.70	39.34	40.10	38.96	39.90	40.35	34.22	40.3	39.1	34.89
0.24	0.64	0.32	0.64	0.21	0.21	0.21	0.10	0.4	0.4	2.00
–	–	–	–	–	–	–	0.01	–	–	–
0.05	0.14	0.03	0.20	0.08	0.05	0.04	0.29	–	–	0.01
none	0.40	0.30	0.24	–	–	–	0.18	–	–	0.09
0.14	–	–	–	–	–	–	0.09	–	–	0.06
1.19	–	–	–	–	–	–	0.10	–	–	0.32
–	–	–	–	–	–	–	<0.001	0.1	0.1	0.10
1.17	1.76	0.96	1.08	–	–	–	13.00	2.5	4.0	2.60
14.38	17.52	14.96	14.82	15.35	15.44	15.54	13.00	14.7	13.4	14.30
–	–	–	–	–	–	–	–	–	–	–
–	–	–	–	–	–	–	–	–	–	0.08
100.35	100.00	100.09	100.18	99.80	100.11	100.49	100.21	101.6	101.3	99.91[a]
195:100	171:100	165:100	180:100	197:100	187:100	187:100	184:100	184:100	192:100	218:100

12. Ross and Kerr (1934): Myeline, Saxony; analyst F.A. Gonyer.
13. Alexander et al. (1943): Djebal Deber, Libya, dried at 110°C; analyst L.T. Alexander.
14. Alexander et al. (1943): Anamosa, Iowa, U.S.A., dried at 110°C; analyst L.T. Alexander.
15. Alexander et al. (1943): Eureka, Utah, U.S.A., dried at 110°C; analyst L.T. Alexander.
16. Swineford et al. (1954): Ness County, Kans., U.S.A.
17. Loughnan (1957): Eureka, Utah, U.S.A.; analyst G.T. See.
18. Loughnan (1957): Bedford, Ind., U.S.A.; analyst G.T. See.
19. Keller et al. (1971): Los Azufres Thermal area, Michoacan, Mexico.

*First 18 from Bates (1959); [a]includes 0.82% SO_4.

clay-rich rocks in general. Ronov and Khlebnikova (1957) reported an average TiO_2 content of 1.15% for 3761 Russian clays.

Most halloysite appears to be the result of supergene processes and, as noted by Ross and Kerr (1934), leaching by sulphuric acid, produced by the alteration of pyrite, appears to be one of the more common weathering processes. Alunite is commonly associated with halloysite (Ross and Kerr, 1934; Swineford et al., 1954). Sulphuric acid is commonly used to dissolve ilmenite in the manufacture of TiO_2 (Willets and Marchetti, 1958). It is possible that in nature it serves the same purpose. In the alteration of granitic rocks and pegmatites, feldspar is more likely to alter to halloysite and mica to kaolinite (Šand, 1956). The micas contain appreciably more titanium than the feldspars which would account for the differences in their alteration products.

Keller et al. (1971) reported on the occurrence of halloysite formed by the action of hot spring waters on rhyolitic volcanic rock in Michoacan, Mexico and suggested that high concentrations of Si and Al in solution, low pH (about 3.5) and sulfate as the solvent anion allows the formation of halloysite rather than other kaolinite minerals.

De Sousa Santos et al. (1965) described a fibrous halloysite from Brasil that showed an appreciable degree of structural regularity in the kaolinite layer structure with b as the fiber axis. The clay was formed by weathering of a porphyritic granite. The chemical analysis of the clay dried at 110°C gives: SiO_2 45.0%, Al_2O_3 40.1%, H_2O (+110°C) 15.4%, Fe_2O_3 0.72%, CaO, MgO, Na_2O trace, K_2O 0.02%, total 101.24%; SiO_2/Al_2O_3 = 191:100. De Sousa Santos et al.(1966) also described a tabular halloysite (or tabular endellite) and concluded that there is no relation between the degree of crystal structure order, the morphology, and the swelling and shrinking characteristics of the layer structure.

A tabular iron-rich halloysite from the Katy soil of Texas was reported by Kunze and Bradley (1964):

	Air dry wt. %	Cation composition
H_2O	16.63	
SiO_2	42.79	4.00
Fe_2O	8.25	0.58
TiO_2	0.88	
Al_2O_3	29.28	3.27
MgO	1.01	0.15
CaO	1.52	0.16
total	100.45	

C.E.C. 60 mequiv./100 g   CaO content 55 mequiv./100 g

This fine-grained clay (primarily less than 0.2 μ in diameter) has an approximate molecular formula of $2SiO_2 \cdot R_2O_3 \cdot 2 \cdot 6H_2O$. The authors believe that because of the high cation exchange capacity and other tests much of the magnesium and ferrous iron is present in the clay structure. The Fe and Mg increase the dimension of octahedral sheet to such an extent that curvature due to a misfit between the octahedral and tetrahedral sheets is not necessary. Mössbauer studies (Weaver et al., 1967) indicate much of the Fe may be present as goethite.

A number of iron-rich halloysites from Russia and one from Italy have been described (Table LXXI). Such clays have been called ferrihalloysite by Vakhrushev and Baxpymeb (1949). These clays are presumably similar to the one described by Kunze

TABLE LXXI

Chemical analyses of halloysites with a high iron content

	1		2		3		4		5		6	
	wt.%	cation comp.	wt.%	cation comp.	wt.%	cation comp.	wt.%	cation comp.	wt.%	cation comp.	wt.%	cation comp.
SiO_2	38.90	4.05	47.40	4.28	45.15	4.74	38.15	3.86	44.43	4.02	41.04	3.95
Al_2O_3	14.83	1.87	25.16	2.64	12.84	1.59	22.04	2.62	29.60	3.16	33.85	3.82
Fe_2O_3	24.50	1.92	9.64	0.65	12.50	0.89	16.95	1.28	8.16	0.56	0.82	0.06
FeO	—		0.98	0.08	nil		0.12	0.01	—		—	
NiO	1.50	0.25	—		—		—		—		abs.	
MgO	trace		1.68	0.23	2.51	0.39	0.73	0.11	1.51	0.20	0.55	0.08
Cr_2O_3	—		—		—		—		—		0.59	0.05
CaO	trace		1.40	0.14	2.30	0.26	3.12	0.34	1.58	0.15	1.24	0.13
Na_2O	—		—		0.15		—		0.26		} 0.30	
K_2O	—		—		0.80		—		0.98			
TiO_2	—		2.72		0.50		—		0.44		abs.	
H_2O^-	10.75		5.81		16.92		7.92		—		10.72	
H_2O^+	10.03	6.66	11.50	6.86	6.64	4.62	10.35	6.95	13.05	7.92	11.01	6.75
Total	100.51		106.29		100.31		99.38		100.51*		100.12	

1. Vakhrushev and Вахрумеb (1949): compact brown ferrihalloysite, Anatolsky, silicate-nickel ore deposit, middle Urals, Russia.
2. Maleev and Maleeb (1949): red ferrihalloysite from weathering of andesite-basalt, volcanic scoria and ash, Amur-Ussuri depression, Russia.
3. Same as no.2 except yellow ferrihalloysite.
4. Efremov (1936): ferrihalloysite, Balka Glubokaia, Russia.
5. Bonatti and Galitelli (1950): dark brown clay occurring as patches and veins in a diatomite bed overlain by volcanic tuff, evidently formed by percolating ground water, Bagnoregia, Italy.
6. Gritzänko and Grum-Grizhmailo (1949): halloysite from Aidyrly deposit, southern Urals.

*includes: 0.03% MnO, 0.47% S.

and Bradley (1964) as a tabular halloysite although the data are not sufficient to show this. The iron in the Russian clays is largely ferric; whereas Kunze and Bradley believe that in the Katy clay it is ferrous. Assuming all the iron is in the octahedral position it fills from 15% to 50% of the occupied postions. It is not known if the high silica and alkali values are due to the presence of impurities. Gritzaenko and Grum-Grzhimailo (1949) reported a light blue chromium halloysite (Table LXXI) and Vakhrushev and Baxpymeb (1949) found 1.50% NiO in a ferrihalloysite.

A number of copper-bearing halloysites have been reported associated with copper deposits (Chukhrov et al., 1970). They believe Cu, equivalent to as much as 7–8% CuO, can occur in the octahedral sheet. When more than this is present, it occurs as chrysocolla, in part in concordant orientation with the halloysite. It would appear that the poorly crystallized kaolinite minerals will allow a much broader range of isomorphous substitution than the better crystallized members.

Bates (1959) studied 64 chemical analyses of 1:1 layer silicates in an attempt to show a relation between composition and morphology (plates or tubes). He noted that halloysite had an average lower SiO_2/Al_2O_3 ratio than kaolinite and dickite and contained more H_2O^+. He concluded that structural misfit due to cation distribution in the octahedral and tetrahedral sheets is probably not the major cause of morphological variation in the kaolinite group. However, he assumed all Fe^{3+}, Mg and excess Al were present in the kaolinite structure and split the distribution between the two different sheets. Neither of these assumptions is likely to be valid. It is not known whether the excess H_2O^+ in halloysite is due to interlayer water or extra H^+ in the structure. In either case, Bates believes the interlayer bonds would be weakened facilitating curling and the formation of tubes. Bates (1959) has suggested that there is continuous morphological series extending from subspherical amorphous allophane to tubular kaolinite, to laths and hexagonal flakes.

Dehydrated halloysites have C.E.C. in the range of 6–10 mequiv./100 g (Van der Marel, 1958; Garrett and Walker, 1959). Garrett and Walker have shown that the exchangable cations are located on the external surfaces of the crystals and not in the interlayer position of halloysite. Until it is possible to obtain accurate chemical analyses of the kaolinite minerals, it will be difficult to determine their exchange capacity and the source of the charge.

Chapter 14

ALLOPHANE

Allophane is a fine-grained amorphous material composed largely of silica, alumina and water. It has both a glassy and an earthy appearance. Material with this composition is clear to white in color; however, because of the common presence of other ions, it can occur in a wide variety of colors. It most often forms as an alteration product and is an abundant component of many soils. Under some conditions it will crystallize into halloysite and halloysite-like clays.

Differential thermal (Belyankin and Ivanova, 1936) and infrared (Adler et al., 1950) studies prove that allophane is not a fine mechanical mixture of alumina and silica but that these are chemically combined as in co-precipitated silica alumina gels. X-ray patterns usually show one or more diffuse bands, which White (1953) interpreted to mean that the structure was more ordered than glass.

Studies by Sudo and Takahashi (1956), Fields (1956), Aomine and Jackson (1959), Robertson (1963), Chukhrov et al., (1963), and others describe the association of allophane with halloysite. Their data indicate that allophane first precipitates as fine particles (0.02 μ) which coagulate into spherical grains. The particles in these grains reorganize into thin ribbons or fibers which further crystallize into fibers and tubes of hydrated halloysite. Tamura and Jackson (1953) proposed a theory in which alumina precipitates and crystallizes to a gibbsite structure. Partial dehydration removes some hydroxyls which are replaced by the oxygen atoms of silica tetrahedra from solution. Silica is attached at random forming a random structure full of channels. When sufficient silica is available, under the influence of wetting and drying, silica reorients to form a hexagonal sheet resulting in a kaolinite-type mineral. This mechanism allows allophane of varying SiO_2/Al_2O_3 ratios with the possibility of progressing towards a kaolinite structure. Bates (1962) has described the reverse process occurring in Hawaii where feldspar alters to halloysite which in turn alters to allophane and eventually gibbsite.

Allophane, in any abundance, is most commonly formed from volcanic material although it can presumably form from any alumino-silicate minerals and indeed is probably present as a transitory stage in the alteration of any material to a clay mineral if any significant structural re-arrangement is required. In its most pure form it occurs as veins and incrustations. Most analyses (Table LXXII) are of samples from this type of deposit. In volcanic soils (Japan, Australia, N.W.U.S., etc.) where allophane is abundant it is usually intimately mixed with halloysite and collection of

TABLE LXXII

Chemical analyses of allophanes

	1	2	3	4	5	6	7
SiO_2	19.71	26.0	26.68	30.23	29.95	31.60	33.22
Al_2O_3	39.71	50.5	44.95	41.81	34.23	32.53	28.24
Fe_2O_3	0.48	1.0	0.12	0.31	0.08	–	0.16
MgO	0.08	–	none	0.04	0.18	0.18	0.14
CaO	–	–	2.37	2.86	1.23	2.46	2.64
P_2O_5	0.52	–	10.57	9.51	–	–	–
SO_3	–	–	0.22	0.44	–	–	0.00
H_2O^-	19.89	–	–	–	22.18	19.26	22.52
H_2O^+	19.78	25.0	16.39	15.61	12.12	13.90	9.80
	100.17[1]	102.5[2]	101.12[3]	100.76[4]	100.25[5]	99.93	100.31[6]
SiO_2/Al_2O_3	84:100	86:100	100:100	121:100	146:100	162:100	197:100

[1] Includes 0.09% Na_2O+K_2O; [2] includes 0.1% TiO_2; [3] H_2O^- = 30.13%; [4] H_2O^- = 34.69%; [5] includes 0.28% CuO; [6] includes 2.07% ZnO, 1.40% PbO, 0.12% F.

1. Michalek and Stoch (1958): white earthy, in veins in silexes, shales and sandstones, Carpathian Flysch.
2. Patterson (1963): weathered of vesicular basalt, Hawaii.
3. White (1953): pink, from Lawrence County, Ind., U.S.A.; analyst L.D. McVicker.
4. White (1953): glassy, from Lawrence County, Ind., U.S.A.; analyst L.D. McVicker.
5. Chukhrov et al. (1963): bluish glassy aggregates on weathered rocks between a pyrite-quartz ore body and an underlying dolomite bed, central Aldan, U.S.S.R.; analyst V.M. Senderova.
6. Chukhrov et al. (1963): colorless glassy incrustations on dolomitized limestone, Podolsk district near Moscow; analyst V.A. Moleva.
7. Chukhrov et al. (1963): glassy incrustations on white porcelain-like zinc clay, Basar-tube in Turkmenia, U.S.S.R.; analyst V.A. Moleva.

pure materials is difficult. The fine-clay content of such soils may contain 20% or more allophane (Aomine and Jackson, 1959). Allophane has been reported in soils from the temperate humid region (Wisconsin) but in minor amounts (De Mumbrum and Chesters, 1964).

When the variety of ions available in a weathering regime and the reactivity of amorphous silica and alumina are considered, most analyses (Table LXXII) of allophane are relatively "pure". Fe_2O_3 and MgO are usually present in amounts less than 0.5%. CuO, ZnO, and PbO are reported under special conditions in amounts ranging from 1 to 4%. CaO is commonly present in amounts up to 3%. Part of the calcium may be present in an exchange position and some may be complexed with the phosphate that is frequently present.

The SiO_2/Al_2O_3 ratios vary from 84:100 (Ross and Kerr, 1934, report a low value of 74:100) to 197:100. Most values are considerably lower than the theoretical value 200:100 of the kaolinite minerals. It appears that when silica and alumina precipitate, alumina is usually in excess of that required for a stable 1:1 structure.

During alteration or crystallization additional positively charged silica must be attached to the negatively charged Si–Al gel. Systematic organization presumably increases with the increase in silica. Chukhrov et al. (1963) heated a number of allophanes to 300°C for 30 hours and noted that only the one with a ratio near 2:1 (197:100) showed any semblance of a 7 Å structure. The structural water content is reasonably constant with molecular ratio H_2O/Al_2O_3, generally being between 2.0 and 2.5. Values are higher than the 2:1 ratio of kaolinites. Adsorbed water is relatively high, ranging from 10 to 25%. This water is presumably adsorbed on the surface and trapped within the gel-like structure.

Ross and Kerr (1934), Jaffe and Sherwood (1950), White (1953), and others have described phosphate-bearing allophane. P_2O_5 values in the 7–10% range appear to be relatively common. The similarity of the silica tetrahedra and the phosphorous tetrahedra would favor such an association. Evansite is an amorphous alumina phosphate and Ross and Kerr (1934) suggested it may occur mixed with allophane. However, it is also possible that the phosphate tetrahedron is a substitute for the silica tetrahedra and occurs as an intimate part of the allophane complex.

Haseman et al. (1951) precipitated phosphates of iron and aluminum and digested them in solutions containing various concentrations of Li, K, NH_4, Cs, Ca, and Mg. They obtained 150 products, most of which were crystalline. Stout (1940) treated halloysite with potassium phosphate and obtained a product similar to taranakite, $H_6(K, NH_4)_3Al_5(PO_4)_8$. Thus, phosphate might be expected to combine readily with allophane and it is likely that it is a much more common component of soil allophanes than has been reported.

White (1953) measured cation exchange capacities of 69.0 and 73.5 mequiv./100 g for Indiana allophanes. Birrell and Gradwell (1956) and Aomine and Jackson (1959) found cation exchange capacity values vary markedly as a function of the type cation, the concentration of leaching solution, the pH of the dispersion reagents, and the amount of washing. Using a method of differences (maximum C.E.C.–C.E.C. after allophane dissolved), Aomine and Jackson obtained an average value of approximately 100 mequiv./100 g.

Allophane occurs in both a spherical and a fibrous form. The limited chemical data are not sufficient to show that morphology is related to composition. Electron micrographs of microtome sections of an Australian allophane (John Brown, personal communication) show that many of the spheres are made up of stacks of thin concentric sheets. Kitagawa (1971) showed high-magnification electron micrographs to demonstrate that the "unit particle" of allophane was spherical and 55 Å in diameter. Yoshinaga and Aomine (1962) described a thread-shaped paracrystalline material with the formula $1.1\ SiO_2 \cdot Al_2O_3 \cdot 2.3-2.8 H_2O(+)$ which they called *imologite*. This material is presumably an intermediate product between allophane and kaolinite.

Chapter 15

TRIOCTAHEDRAL 1:1 CLAY MINERALS

Many of the trioctahedral 1:1 minerals that have been studied are not considered to be clay minerals; however, it is likely that minerals of similar composition and structure exist as clay-sized material but are seldom concentrated enough to be analyzed or even identified.

The serpentines are reviewed briefly and the other trioctahedral 1:1 clays, which more commonly occur as clay-sized minerals, will be discussed in some detail. Faust and Fahey (1962) and Deer et al. (1962) have written excellent review articles on the serpentine subgroup.

Most serpentines deviate very little from the ideal composition $Mg_6Si_4O_{10}(OH)_8$. The calculated structural formulas for most samples show little or no tetrahedral substitution, although Faust and Fahey show values as large as 0.41 Al^{3+} and 0.26 Fe^{3+}. Al^{3+}, Fe^{3+}, and Fe^{2+} commonly substitute for Mg^{2+} in the octahedral sheet. For the 28 selected analyses given by Faust and Fahey octahedral Al^{3+} ranges from absent to 0.30, averaging 0.05; octahedral Fe^{3+} absent to 0.37, averaging 0.09; octahedral Fe^{2+} absent to 0.45, averaging 0.09. The serpentine minerals formed by the reaction of hydrothermal solutions with ultrabasic rocks have a much higher iron and aluminum content than those formed in carbonate rocks. For these same samples the sum of the tetrahedral cations range from 4.00 to 4.10 and octahedral cations from 5.80 to 6.04. The antigorite samples have only 5.80 to 5.84 of the octahedral positions filled. The rectified wave structure proposed by Zussman (1954) for antigorite requires that there be a deficiency of octahedral cations and anions with respect to the tetrahedral sheet.

It has been suggested that the morphology is controlled by the relative size of the octahedral and tetrahedral sheets (Bates, 1959; Deer et al., 1962); the theoretical octahedral sheet is considerably wider than the theoretical tetrahedral sheet. The substitution of Al^{3+} tends to increase the size of the tetrahedral sheet and decrease the size of the octahedral sheet. The platy antigorites tend to contain more Al^{3+} than chrysotile and lizardite. The sheet misfit, in part due to the low Al^{3+} content, is presumably responsible for the tubular morphology of chrysotile and the fine size of lizardite.

Bailey and Tyler (1960) have found numerous examples of aluminous serpentines associated with the iron ores from Lake Superior in Michigan. Basal X-ray spacings indicate various compositions, ranging from $(Si_{3.5}Al_{0.5})(Mg_{5.5}Al_{0.5})O_{10}(OH)_8$ to

$(Si_{2.25}Al_{1.75})(Mg_{4.25}Al_{1.75})O_{10}(OH)_8$. They noted four different types of X-ray patterns which they explained on the basis of mixtures of Al-lizardite with two different 6-layer orthohexagonal structures. The proportions and compositions of each of these structural types were variable in any one specimen.

Serpentine-like minerals (1:1 trioctahedral) with relatively large amounts of isomorphous substitution have been given the names: chamosite, berthierine, amesite, cronstedtite, and greenalite. Early studies suggested that some of these minerals had a 14 Å chlorite structure; however, more recent data indicate most have a 7 Å serpentine-like structure (Brindley, 1961b). The Nomenclature Committee (Bailey et al., 1971) decided that the name "berthierine" had priority over "chamosite"; however, it does not seem to us prudent to make such a change at this time. These minerals have chemical compositions similar to the chlorites and as shown by Nelson and Roy (1958) some of them can be transformed into 14 Å chlorites at temperatures higher than 400°–500°C. For this reason they have suggested the name *septechlorite* for these minerals.

Chamosite appears to be the finest grained and most abundant mineral in this group. It occurs in lateritic clay deposits (Brindley, 1951), both as oolites and matrix in sedimentary ironstones (Hallimond, 1925), in hydrothermal deposits (Ruotsala et al., 1964), in shales (Drennan, 1963), in Recent shallow-marine deposits (Porrenga, 1966) and in estuarine sediments (Rohrlich et al., 1969). It is probable that chamosite is more abundant than commonly realized; however, Drennan (1963) has pointed out that it is extremely unstable in a leached and oxidized environment and is not likely to persist as an allogenic mineral.

The chamosites in individual sedimentary iron formations vary widely in color which indicates there is considerable local variation in chemical composition. The sedimentary chamosites are usually quite impure, frequently containing siderite and goethite. It is necessary to make a number of assumptions in recasting analyses of these samples.

Chamosites have appreciable tetrahedral and octahedral Al^{3+}. The following general formula was suggested by Brindley (1961b):

$(Fe^{2+}, Mg)_{2.3}(Fe^{3+}, Al)_{0.7}(Si_{1.4}Al_{0.6})O_5(OH)_4$

The wide variation shown by chemical analyses (Table LXXIII) indicates that this is at best a typical formula. Structural formulas are given in Table LXXIV.

Tetrahedral Al^{3+} values range from 0.10 to 0.76 per two tetrahedral positions and average 0.49 for the ten selected samples. The amount of tetrahedral substitution of the larger Al^{3+} for the smaller Si^{4+} is appreciably greater than has been reported for the 2:1 clays, although the higher values are in the range (0.9–1.5 per four tetrahedral positions) reported for the trioctahedral micas (Foster, 1960) and chlorites (Brown and Bailey, 1962).

TABLE LXXIII

Chemical analyses of some chamosites

	1	2	3	4	5	7	9	10
SiO_2	22.03	23.81	26.40	25.5	24.50	24.65	38.3	32.6
Al_2O_3	22.91	23.12	18.23	14.9	16.50	17.55	14.6	10.25
Fe_2O_3	0.49	0.23	5.70	–	7.45	19.89	15.1	38.9
FeO	36.68	39.45	25.87	35.6	31.46	12.58	8.8	n.d.
MgO	1.91	2.72	11.35	5.6	4.59	5.70	10.5	4.79
MnO	0.04	0.04	–	1.0	3.33	–	(0.36)[b]	1.71
CaO	0.07	–	0.42	–	–	0.90	(0.86)	(0.72)[c]
Na_2O	0.08	–	0.17	–	–	–	(0.18)	–
K_2O	0.03	–	0.17	–	–	–	(0.59)	(1.30)
TiO_2	3.63	–	–	–	–	–	(0.32)	0.23
H_2O^+	10.65	10.67	10.60	17.4	11.36	14.32	12.2	–
H_2O^-	0.63	–	1.05		–	4.41	–	–
Total	100.05[a]	100.00	100.00	100.0	99.01	99.97	100.00	100.00

[a] includes: $P_2O_5 = 0.18$, $CO_2 = 0.04$, $SO_3 = 0.27$, $Cr_2O_3 = 0.05$, organic = 0.03; [b] present in uncorrected samples, also, 0.65% P_2O_5; [c] present in uncorrected samples, also, 0.81% P_2O_5.

The composition of the octahedral sheet of most chamosites is similar to that of the siderophyllites and lepidomelanes (Fig.24) as reported by Foster (1960). The average values for the octahedral sheet (allowing for non-determined Fe^{2+}) is:

$$(Al_{0.69}Fe^{3+}_{0.24}Fe^{2+}_{1.35}Mg_{0.43})_{2.71}$$

Octahedral Al ranges from 0.30 to 1.01. The ratio of octahedral Al to tetrahedral Al ranges from 1.0 to 7.3 with most values being near 1.0. Fe^{2+} is the dominant octahedral cation ranging from 0.36 to 1.84, but in most samples occupies more than 50% of the filled octahedral positions. Fe^{3+} values range from 0.0 to 0.60 and to a large extent may be a function of post-formation oxidation. Mg ranges from 0.16 to 0.91 with most values being less than 0.50.

Ruotsala et al. (1964) reported a partial analysis of a chamosite which contained 18.7% MgO and only 9.70% FeO. This is enough Mg to fill approximately half the octahedral positions and Fe^{2+} to fill one-seventh. If this analysis is valid, the composition of the octahedral sheet would be similar to that of Mg-biotites (Fig.24), considerably extending the range of isomorphous substitution. If chamosite behaves as the trioctahedral micas (Foster, 1960), then as octahedral Mg increases at the expense of Fe^{2+}, octahedral Al and tetrahedral Al both tend to decrease and the composition

TABLE LXXIV

Structural formulas of some chamosites calculated on basis 5 (O) and 4 (OH)

	1	2	3	4	5	6	7	8	9	10
Tetrahedral										
Si	1.24	1.33	1.42	1.54	1.57	1.59	1.41	1.36	1.90	1.68
Al	0.76	0.67	0.58	0.46	0.43	0.41	0.59	0.64	0.10	0.32
Octahedral										
Ti	0.15	–	–	–	–	–	–	0.01	–	0.01
Al	0.76	0.85	0.60	0.59	0.63	1.09	0.59	0.81	0.73	0.30
Fe^{3+}	0.02	0.01	0.23	–	0.21	0.12	0.60	–	0.55	1.50
Fe^{2+}	1.73	1.84	1.13	1.79	1.45	1.30	0.85	1.70	0.36	n.d.
Mg	0.16	0.22	0.91	0.50	0.37	0.16	0.49	0.36	0.77	0.37
Mn	–	–	–	0.05	0.15	–	–	0.04	–	0.07
	2.82	2.92	2.87	2.93	2.81	2.67	2.57*	2.91	2.41	2.25
Tetrahedral charge (–)	0.76	0.67	0.58	0.46	0.43	0.41	0.59	0.64	0.10	0.32
Octahedral charge (+)	0.72	0.72	0.57	0.45	0.46	0.55	0.37	0.63	0.10	0.32

*Includes 0.06 Ca.

1. Brindley (1951): hard greenish-gray clay derived from alteration of olivine-basalt by sub-aerial weathering, Ayrshire coal fields, area III, England; analyst B.E. Dixon.
2. Brindley and Youell (1953; analysis recast to 100.00% by Deer et al., 1962): in matrix of siderite, Stanion Lane pit, Corby, Northamptonshire, England; analyst: R.F. Youell.
3. Banister and Whittard (1945): oolites from Silurian siltstone, Wickwar, England; high Mg content is believed to be due to post-depositional Mg-rich solutions which dolomitized enclosing limestone (14 Å variety).
4. Deudon (1955): from Meurth et Moselle, France; analysis recast to 100%.
5. Sudo (1943): spherical or cylindrical aggregates in quartz veins and in irregular masses of chalcopyrite from Arakawa mine, Akita, Japan.
6. Youell (1958): from chamosite-kaolinite beds of Northampton sand ironstone formation at Wellingborough, England.
7. Novak and Valcha (1964): in drusy cavities of hydrothermal polymetallic veins as the latest mineral, Hora Svate Katering in the Krusne Hory Mountains, Czechoslovakia.
8. Schoen (1964): black oolites from Clinton iron stone bed near Clinton, N.Y., U.S.A., corrected for impurities.
9. Schellmann (1966): goethite grains altered in Recent shallow-marine environment, Guinea coastal sands; corrected for 24% goethite and recast to 100%.
10. Rohrlich et al. (1969): pellets formed in Recent muds, Loch Etive, Scotland; corrected for impurities and recast to 100%.

TABLE LXXV

Chemical analyses and structural formulas of amesite, greenalite, and cronstedtite

	1	2	3	4	5
SiO_2	20.95	30.08	38.00	35.92	16.42
Al_2O_3	35.21	–	–	–	0.90
Fe_2O_3	–	34.85	8.40	9.80	29.72
FeO	8.28	25.72	46.56	46.16	41.86
MgO	22.88	–	–	–	–
CaO	0.58	–	–	–	1.32
H_2O^+	13.02	9.35	7.09	8.11	10.17
H_2O^-	0.23	–	–	–	–
Total	101.15	100.00*	100.00*	100.00*	100.39
Octahedral					
Al	1.0	–	–	–	–
Fe^{3+}	–	1.23	0.35	0.42	0.71
Fe^{2+}	0.33	1.29	2.20	2.22	2.38
MgO	1.64	–	–	–	–
	2.97	2.52	2.55	2.64	3.09
Tetrahedral					
Si	1.01	1.74	2.14	2.07	1.12
Al	0.99	–	–	–	0.07
Fe^{3+}	–	0.26	–	–	0.81

1. Gruner (1944): amesite, pale bluish-green crystals, Chester, Mass., U.S.A.; analyst F.V. Shannon.
2. Gruner (1936): greenalite, no. 45758, granular aggregates, recalculated to 100.00%, Mesabi Range, Minn., U.S.A.
3. Gruner (1936): greenalite, no. 45766.
4. Gruner (1936): greenalite, Follife analysis.
5. Hendricks (1939b): cronstedtite, Kisbanya, Hungary; analyst Gossner.

*Some MgO is probably present.

approaches that of serpentine. Because Mg is smaller than Fe^{2+}, less Al is required to adjust the octahedral and tetrahedral sheet sizes.

The filled octahedral positions are less than 3.00, ranging from 2.25 to 2.93. Except for the two Recent samples all values are larger than 2.50 (similar to trioctahedral micas) and most are larger than 2.80. Total R^{3+} ions range between 20–53% of the total octahedral cations. The octahedral charge ranges from + 0.10 to + 0.72 with an average value of + 0.49 (the eight older samples have + 0.56 and the two Recent + 0.22). All analyses examined showed both a positive octahedral charge and a deficiency in octahedral position. Thus, the positive charges created by the trivalent cations proxying for bivalent cations are accommodated both by a negative tetrahedral charge and by negative charges due to unoccupied octahedral sites. In most instances, the former is the major source of negative charge. The positive octahedral charge is closely equivalent to the negative tetrahedral charge.

In serpentines with a composition of $Mg_6Si_4O_{18}(OH)_8$ the octahedral sheet is appreciably larger than the tetrahedral sheet. The resulting stress is in part relieved by bending. Gillery (1959) has synthesized a range of minerals with the general formula $(Si_{4-x}Al_x)(Mg_{6-x}Al_x)O_{10}(OH)_8$. Al is presumed to be equally divided between the two sheets in order to maintain neutrality. As the Al content is increased, the tetrahedral sheet becomes wider and the octahedral narrower, decreasing the interlayer strain. With a sufficiently high Al content ($x > 0.75$) the tetrahedral sheet becomes larger than the octahedral and size adjustment is made by tetrahedral rotation as in the dioctahedral clays (Radoslovich, 1963). When $x = 0$ a fibrous serpentine was formed. When $x = 0.25$ and more, the strain was small enough that the serpentine had a platy morphology.

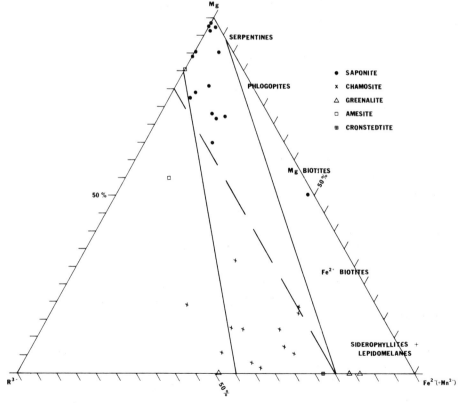

Fig. 24 Composition of the octahedral sheets of the trioctahedral clay minerals. The two solid boundary lines are those established by Foster (1960) for natural biotites. Dashed line separates 2:1 and 1:1 clays.

Although the octahedral sheet of chamosite is composed predominantly of Fe^{2+} (radius = 0.75) rather than the smaller Mg (radius = 0.65), the Al content ($x > 1.20$) is sufficiently large so that the tetrahedral rotation is necessary to adjust the size differential (Radoslovich, 1963). As Mg increases relative to Fe^{2+}, less Al is required to afford a strain-free structure. There appears to be no reason that there cannot be a continuous isomorphous series between chamosite and serpentine.

Low-temperature sheet structure silicates with a high Fe^{2+} content seem to be restricted to the 1:1 minerals. The Fe^{2+} content of the octahedral sheet of the 2:1 clays is seldom larger than 0.6 per 3.0 positions (less than 0.3 for most samples). Few low-temperature 2:1 clays have enough Al substitution in the tetrahedral sheet to adjust its size to that of the large octahedral sheet. Substitutions of this magnitude, 1 Al per four tetrahedral positions, at low temperatures, are favored more by the 1:1 than the 2:1 arrangement. Presumably, the lack of a layer charge and, therefore, the need for the tetrahedra to rotate to accommodate an interlayer K, and the fact that the tetrahedral sheet is sandwiched between two octahedral sheets allow interlayer size adjustments to be made more easily in the 1:1 than in the 2:1 clays.

Powder diagrams of 7 Å chamosites vary considerably. Brindley (1951) has shown that this could be explained by the presence of both orthogonal and monoclinic stacking of layers. These two types occur in varying proportions in different specimens though the orthogonal form generally predominates.

Random displacements of $nb/3$ parallel to b are present in chamosite as in kaolinite (Brindley, 1961). Youell (1955) has shown that in many chamosites there is disorder parallel to the a-axis. The better ordered chamosites are rich in iron, and the less well ordered contain less iron and more aluminum. Oxidation of Fe^{2+} to Fe^{3+} also increases the disorder. Both mechanisms cause a larger ion to be replaced by a smaller. This results in a decrease in the size of the octahedral sheet thus decreasing the misfit between the octahedral and tetrahedral sheets. Decreasing the strain within the layers apparently allows more shifting between layers. Youell (1955) noted that "disordered" chamosite is "almost invariably accompanied by kaolinite, and ordered chamosite by siderite".

Chamosites are formed in a variety of ways but the chemical conditions must be fairly restricted. Porrenga (1966) found chamosite faecal pellets being formed in the organic muds of the Niger Delta from detrital iron-silicate mixtures and Schellmann (1966) reported the formation of chamosite from detrital goethite grains in the shallow marine sands off the coast of Guinea. Chamosite has also been found forming in the relatively fresh water environment of a restricted portion of Loch Etive, Scotland (Rohrlich et al., 1969). Thus, the presence of organic material, detrital iron, and a slightly basic pH appear to be the basic environmental requirements.

It has been suggested by some that chamosites suffer late diagenetic modifications (see Rohrlich et al., 1969). Analyses of Recent samples are not too reliable but suggest that recently formed chamosites have a lower octahedral population, largely Fe^{2+}, and less tetrahedral Al than older samples. This suggests that these two ions may continue to be added to the clay layer during burial. The data do not indicate that any appreciable number of cations are released from the chamosite, with the possible exception of Si.

Amesite has more Al substitution in the tetrahedral and octahedral sheets than has chamosite (Table LXXV). In addition, the octahedral sheet is composed predominantly of Mg ions rather than Fe^{2+}. Some samples show considerable disorder due to random displacement of layers by multiples of $b/3$ parallel to the y-axis (Deer et al., 1962). In both amesite and chamosite the negatively charged tetrahedral sheets and positively charged octahedral sheets allow ionic bonding between adjacent layers and a resulting contraction normal to the c-axis. The layer thickness is on the order of 7.00–7.11 as compared to 7.15 for kaolinite. This mineral is quite rare and has not been found in sedimentary deposits.

Greenalite and cronstedtite are also quite rare; in both, Fe^{2+} is the predominant cation (Table LXXV). In greenalite the octahedral sheet is composed largely of Fe^{2+} with a minor amount of Mg. There is only limited substitution by Fe^{3+} in the tetrahedral sheet. This mineral is difficult to obtain in a pure form and all analyses require calculated corrections. The Fe^{3+}/Fe^{2+} ratio varies largely as a function of weathering. Greenalite has been found only in the iron sediments of the Mesabi Range in Minnesota. It occurs in a granular form and is apparently a direct precipitate. The special set of conditions that favor the formation of this mineral are not understood but Gruner (1936) suggests the almost toal absence of Ca, Al, and alkalies may have been a factor.

Chapter 16

LOW-TEMPERATURE SYNTHESIS

Considerable work has been done on the low-temperature (below 100°C) synthesis of the clay minerals and all the common clays except attapulgite have been synthesized. In spite of this, the processes involved are not well understood. Sedletsky (1937), Caillère and Hénin (1950), Caillère et al. (1953), Caillère (1953, 1954), Hénin and Robichet (1953) and Hénin (1954,1956) have pioneered in the low-temperature synthesis of clay minerals.

Hénin (1956) has summarized the results of the early French studies. Co-precipitation of silica and alumina or iron produced amorphous products which could not be crystallized by mild heating or by alternate wetting and drying. In an attempt to facilitate organization of the constituents they tried an electrolytic method and were able to produce material which resembled antigorite but had a deficiency of silica. They apparently formed hydroxide layers which attracted minor amounts of silica. Their best success was obtained using very dilute solutions (several tenths of a milligram per liter). "Stable alkaline solutions of silicates or aluminates are placed in one flask and stable acid solutions in another flask. These two solutions are allowed to flow at a rate of several cc per day into a large glass vessel of 5 l capacity containing 2 l of distilled water". They were able to make Mg, Fe^{2+}, Fe^{3+}, Zn, and Ni montmorillonites. By adding KCl to the flasks they were able to form a mica type clay. The phlogopite variety was the most stable. Using Mg or Ni solutions in place of distilled water, they were able to form Mg and Ni antigorites.

The crystallinity was improved by the presence of such electrolytes as NaCl and $CaSO_4$ even though they did not enter into the reaction. The role of these electrolytes is not understood. Frank and Evans (1945), Bradley and Krause (1957) and others have shown that large anions such as Cl^- and SO_4^{2-} tend to breakdown the water structure. This would presumably increase the mobility of the small clay building cations (which are water structure-formers) and facilitate their organization.

If sufficient silica was present, the ratio of silicate to cation had little effect on the nature of the substances obtained. If the solution was deficient in silica, additional silica was obtained from the walls of the flask. In some instances, the cations combined with the glass to form clays on the walls of the flask. Following this lead, they were able to place basalt in contact with a magnesium solution and form montmorillonite. The silica content of products formed is larger in a neutral or alkaline medium than in an acid one. At low pH values, the silica sheets are incomplete and the hydroxide sheet is continuous. At higher pH values, the silica layers are continuous and the hydroxide layers discontinuous.

It was easier to prepare clays using cations that formed a brucite structure rather than those forming a gibbsite structure. This is, in part, due to the difficulty of forming octahedrally coordinated Al^{3+} compounds in the presence of silicon at other than low pH conditions. Hénin and co-workers believed that hydroxide sheets are formed on which silica is subsequently fixed. The presence of silica tends to cause the hydroxides to precipitate at lower pH than they normally do. Most of these experiments were carried on at 100°C, although crystalline products were obtained at temperatures of 20°C. The rate of formation varies in the ratio 1:560 between 0°C and 100°C.

Iler (1955) has provided the following explanation of some of the difficulties involved in precipitating crystalline silicate minerals. Most solutions of soluble silicates consist of a mixture of polysilicate ions, containing up to 5 or 10 SiO_2 units each. Highly alkaline metasilicates contain monomeric silicate ions but upon lowering the pH by the addition of a solution of a metal salt rapid polymerization occurs. The polysilicate ions are not of uniform size and can not arrange themselves along with the metal ions into a regular crystal structure. Thus the precipitation of polysilicate ions with metal ions almost always results in an amorphous complex. The polysilicate ions must be depolymerized to smaller silicate ions of uniform size in order to rearrange themselves into a regular structure.

The polymerization of basic metal ions will also interfere with crystallization. When a metal salt is mixed with a silicate, the increase in pH in the environment of the metal ion causes the formation of polymeric basic metal ions of colloidal metal hydroxides which are not likely to fit into a silicate crystal. Thus, the precipitation of a metal silicate from aqueous solution at normal temperature tends to produce a coagulation of positively charged colloidal metal hydroxide and negatively charged colloidal silica.

Iler also noted that in dilute solutions colloidal metal silicates precipitate at a pH slightly below that at which the metal hydroxide alone would be precipitated. The strong tendency for some hydroxides to react with silica is demonstrated by the fact that the addition of 300 p.p.m. of $Mg(OH)_2$ to water will reduce the soluble silica content from 42 to 0.1 p.p.m. Aluminum oxide is capable of reducing the solubility of silica from 170 p.p.m. to 20 p.p.m. Cations capable of forming insoluble silicates will reduce the solubility of amorphous silica. Al^{3+} at a concentration of 100 p.p.m. reduces the solubility of silica from 120 p.p.m. to 1 p.p.m. at pH 8—9 (Okamoto et al., 1957).

Siffert (1962) reviewed the literature on low-temperature synthesis and reported the results of his own studies. In a saturated solution of silica with a SiO_2/MgO mole ratio of 0.702 he was able to crystallize a trioctahedral montmorillonite at pH 11.3. Using a mole ratio of 1.432 he obtained talc at pH 12. From a SiO_2/MgO mole ratio of

0.70 sepiolite formed at pH 8. At pH values above 9 the sepiolite disappeared and montmorillonite or serpentine formed. The SiO_2/MgO ratio of the precipitate systematically decreased as the pH increased from 8.25 to 11.51.

Dioctahedral Al clays are more difficult to make than trioctahedral Mg clays largely because of the difficulty of forming octahedrally coordinated Al. Cations with a radius larger than 0.62 Å readily enter into octahedral coordination; Al with a radius of 0.50 Å enters into six-fold coordination with difficulty. In an effort to overcome this problem, Siffert (1962) used Na $Al(C_2O_4)_3$ in which the Al is in six-fold coordination. He was able to form a precipitate which afforded a broad 7 Å X-ray reflection. Chemical analyses showed the SiO_2/Al_2O_3 ratio was appreciably lower than that of kaolinite. The best crystallized material was formed in a pH range of 7.25 to 7.75 and as the SiO_2/Al_2O_3 ratio in the starting solution decreased, the pH at which the best crystallized products formed decreased. It was suggested that in nature humic material serves a function similar to that of the oxalic anion.

DeKimpe et al. (1961) conducted a series of experiments in an attempt to synthesize kaolinite. They noted that with increasing pH there was an increasing tendency for Al to substitute for Si in the precipitated gel. This resulted in an increased C.E.C. There was a tendency for Al in four-fold coordination to decrease relative to that in six-fold coordination as pH decreases, thus at low pH (4.5—5.0) they were able to synthesize minor amounts of kaolinite. At a higher pH a mica-like clay was formed.

In a later study (DeKimpe et al., 1964) they demonstrated that it was necessary that the Al be in six-fold coordination before it would combine with Si to form kaolinite. Starting with the Al in six-fold coordination presumably decreased the free-energy barrier necessary for the formation of kaolinite. Gibbsite, $Al(OH)_3$, was used as a starting material but was too stable to be attacked by depolymerized silica; it was necessary for the gibbsite to be in the process of reorganizing ("dynamic" stage) into boehmite, AlO(OH).

Since these studies were conducted, efforts have been directed toward synthesizing kaolinite at lower temperatures (Poncelet and Brindley, 1967; De Kimpe and Fripiat, 1968; De Kimpe, 1969; Eberl, 1970). Kittrick (1970) and Linares and Huertas (1971) reported some success at synthesizing kaolinite at room temperature. Kittrick used three different naturally occurring montmorillonites as "solid-phase aluminum-silicate ion sources". The sample solutions were analyzed at least six times each over a period from three to four years at the end of which time two of the three montmorillonites in solutions saturated to slightly supersaturated with respect to kaolinite had varying amounts of poorly crystalline kaolinite present. Linares and Huertas (1971) used fulvic acid derived from peat as a means of getting Al into six-fold coordination. Variables in the fulvic-acid Al solutions were the SiO_2/Al_2O_3 ratios and the pH (4—9). The precipitation products aged for one month at room temperature. Poorly crystalline kaolinite or "prekaolinite" was synthesized at SiO_2/Al_2O_3 molar ratios from 1 to 10 throughout the pH range of the experiment. Gibbsite, bayerite, boehmite, and silico-alumina gels were also formed. (Gibbsite formed when the molar ratio of SiO_2/Al_2O_3 was low and the solution acid.) Linares and Huertas proposed the following as a mechanism for the formation of kaolinite:

"When the pH of the ternary system consisting of $Si(OH)_4$, the aluminum fulvic acid complex, and H_2O in acid medium is changed, colloidal aluminum hydroxide begins to precipitate slowly. This hydroxide may be formed by a competitive reaction among the organic ligands and the hydroxyl ions leading to a pregibbsitic structure; later, by a surface reaction, monomeric silica can be adsorbed. In this way, a tetrahedral layer of silica is 'soldered' over an octahedral layer of aluminum, with the result that a 1:1 clay mineral, kaolinite, is formed."

Gastuche (1963) reviewed the recent work she and others have done on the synthesis of the octahedral sheet. It is generally thought that the octahedral sheet acts as a template on which the silica framework somehow develops. Al, Mg, and Fe octahedral hydroxides are readily formed and fairly well understood. A number of mixed or double hydroxides (Al-Mg, Al-Fe, etc.) have been formed (Feitchnecht, 1962) but relatively little work has been done on them.

Chapter 17

RELATIONS OF COMPOSITION TO STRUCTURE

Trioctahedral sheet

The distribution of the cations in the octahedral sheet of both the 1:1 and 2:1 trioctahedral clay minerals is shown in Fig.24. The solid boundary lines are those established by Foster (1960) for biotite.

The make-up of the octahedral sheet and the structure are closely related. The high-Mg samples all have a 2:1 structure and the high-Fe^{2+} samples a 1:1 structure. This presumably reflects the difficulty the two silica sheets have in adjusting to a large Fe-octahedral sheet. The single silica layer of the 1:1 clays requires a large amount of Al substitution to adjust to the large octahedral sheet. This results in an increase in charge which is compensated, in the 1:1 clays, by substitution of Al^{3+} and Fe^{3+} in the octahedral sheet. This latter fact is illustrated in Fig.24 where it can be seen that most of the 1:1 clays have more than 20% octahedral occupancy by R^{3+} ions and all 2:1 clays have less than 20% occupancy.

If the silica sheets in a 2:1 Fe-rich clay had enough Al substitution to adjust to the size of the octahedral sheet, this would most likely provide sufficient layer charge to cause contraction and the formation of a biotite mica, or the bonding of a positively charged brucite sheet and the formation of a chlorite. Low-temperature micas of this composition may well exist, but have not yet been recognized. It is not known if low-temperature chlorites of this composition can form.

Several of the iron-rich clays as well as an amesite have a larger R^{3+} value than that allowed for the biotites. Nelson and Roy (1958) were not able to synthesize a 1:1 (amesite) or a 2:2 (chlorite) mineral with more than 33% Al substituting for Mg ($Mg_8 Al_4$) in the octahedral sheet. However, if the analyses in Fig. 24 are correct, it appears that when Fe^{2+} rather than the smaller Mg is the dominant cation in the octahedral sheet, up to 50% R^{3+} can occur in the octahedral position. Foster's (1960) biotite data show a similar relation.

With increased R^{3+} substitution, the positive charge in the octahedral sheet increases. This would increase cation-cation repulsion. The larger the cations the smaller would be the structural strain imposed by this repulsion. In addition, as the R^{3+} content of the octahedral sheet is increased, it is matched by an equivalent amount of R^{3+} in the tetrahedral sheet. As a result of R^{3+} substitution, the tetrahedral sheet increases in size and the octahedral sheet decreases. Eventually the tetrahedral sheet

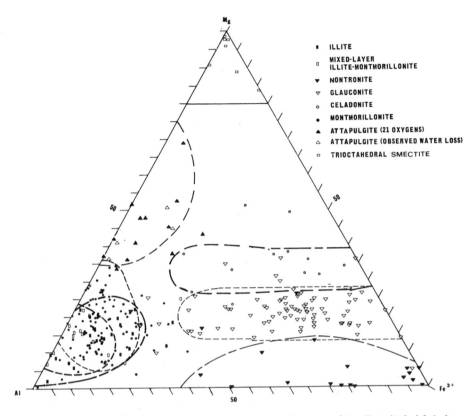

Fig.25. Ternary plot based on the composition of the octahedral sheet of the dioctahedral 2:1 clay minerals, attapulgite and trioctahedral saponite. Fe^{2+} was excluded.

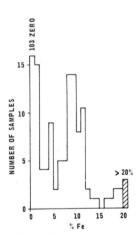

Fig.26. Histogram showing the distribution of the percent of Fe^{2+} in the octahedral sheet of the dioctahedral 2:1 clay minerals. No Fe^{2+} was reported for 103 samples and three samples had more than 20% Fe^{2+}.

becomes larger than the octahedral and with further increase in R^{3+} the tetrahedra must rotate to adjust to the size of the octahedral sheet. Because the Fe^{2+}-rich octahedral sheet is appreciably larger than the Mg-rich octahedral sheet, both the octahedral and tetrahedral sheets can accommodate more R^{3+} substitution before the tetrahedra are forced into excessive rotation.

The chlorite clays appear to have octahedral sheets with compositions that are largely intermediate between the 2:1 and 1:1 octahedral sheets. The chloritic structure allows for a wider range of substitution than the other clays. In part this is because most data on the octahedral composition are an average of two octahedral sheets, each of which could have relatively restricted compositions.

Dioctahedral sheet

Octahedral Mg, Al, and Fe^{3+} for the dioctahedral clays were totalled and the relative proportion calculated (atomic percent). Octahedral Mg, Al, Fe^{3+} and Fe^{2+} were totalled and the percent Fe^{2+} calculated. The distribution values for the 2:1 minerals are summarized in Fig.25. For comparison purposes, values for attapulgite and some trioctahedral smectites are also shown.

There is almost complete coverage of the lower portion (less than 40% Mg) of the diagrams. The high-Fe^{3+} octahedral sheets have a higher maximum amount of Mg (50%) than the high-Al sheets (35%). The Fe^{3+}-rich sheets also have a higher Fe^{2+} content than the Al-rich sheets and, in general, the Fe^{2+} content increases as the Mg content increases. The average octahedral Fe^{2+} for the clays is: illite 1.7%, mixed-layer illite-montmorillonite 0.6%, montmorillonite 0.9%, nontronite 0.7%, glauconite 9.9%, celadonite 10.0%. Although there is considerable error in the Fe^{2+} values, the relative differences are real. If the Fe^{2+} values were added to Mg values, the celadonite and glauconite fields would move appreciably closer to a Mg + Fe^{2+} apex.

The maximum amount of Fe^{2+} the octahedral sheets of the 2:1 dioctahedral clays generally contain is 12% (Fig.26). This is equivalent to approximately 0.25–0.30 octahedral positions. Due to the relatively large size of the Fe^{2+} ion, the structure can apparently adjust to only about half as much Fe^{2+} as Mg.

When the modal values are considered, there is relatively little overlap in the composition of the octahedral sheets of the three Fe^{3+}-rich clays; whereas, for the Al-rich clays, the illite and mixed-layer illite-montmorillonite fields fall within the montmorillonite-beidellite field.

Nomenclature is slanted in favor of Fe over Al. When the octahedral sheet has more than 30% Fe^{3+}, the clay is usually given a Fe-mineral rather than an Al-mineral name. The data indicate that other than a restriction of less than 40% Mg all combinations exist in nature although most of the values are clustered in two areas: greater than 60% Al and greater than 40% Fe^{3+}. As long as the Mg and Fe^{2+} content of the octahedral sheet is below the limits stated, the tetrahedral sheet is able to adjust to the size and charge of the octahedral sheet by cation substitution and tetrahedral twisting and can form a stable structure.

Although some tetrahedral-octahedral combinations are more stable than others, the relative abundance of certain octahedral compositions (Fig.25) may be in part due to the restrictions imposed by the major natural environments. For example, the environmental conditions (and source material) under which most clays form is such that either Al or Fe may be relatively abundant, but seldom both. The plot also shows that the octahedral sheet of attapulgite tends to be composed of approximately 50% Mg and 50% Al and their compositional field does not overlap that of the 2:1 dioctahedral and trioctahedral clays.

Dioctahedral and trioctahedral sheets

Fig.27 shows the compositions of the octahedral sheets of the low-temperature dioctahedral and trioctahedral clay minerals calculated in the basis of R^{3+}, Mg, Fe^{2+} +

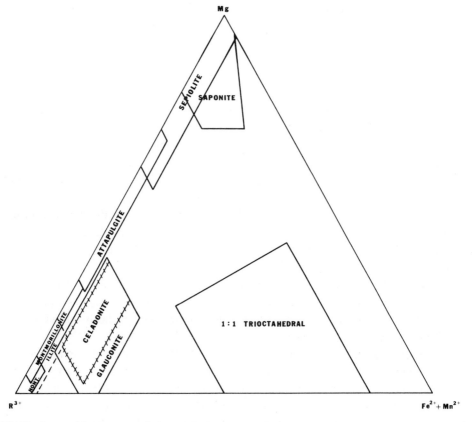

Fig.27. Compositional range of the octahedral sheet of the low-temperature dioctahedral and trioctahedral clay minerals.

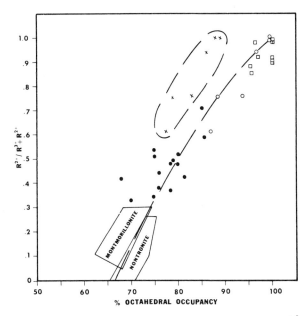

Fig.28. The relation of percent octahedral occupancy to $R^{2+}/(R^{3+}+R^{2+})$ for layer structure and chain structure clays. □ = saponite; ● = attapulgite; x = sepiolite (nine octahedral positions); o = sepiolite (eight octahedral positions).

Mn^{2+}. The clay minerals tend to be concentrated in three relatively discrete areas representing Mg-rich sheets, R^{3+} (both Al and Fe^{3+})-rich sheets and Fe^{2+}-rich sheets. Due to the large size of the Fe^{2+}, octahedral sheets containing only this ion do not normally occur in the clays.

Attapulgite-palygorskite clays lie between the Mg-rich and R^{3+}-rich sheet clays. Thus, a continuous series exists between Mg and R^{3+}. The Mg/R^{3+} ratio in this series is related to the relative amount of filled octahedral positions per total positions available. The dioctahedral clays theoretically have 67% of their total octahedral positions filled; the attapulgites 80%; the sepiolites 90% and the trioctahedral clays approximately 100%. Fig.28 shows a plot of percent octahedral occupancy versus $R^{2+}/R^{3+} + R^{2+}$ illustrating this relation. As R^{2+} is replaced by R^{3+} there is a gradual increase in unoccupied positions to satisfy charge requirements.

The dashed line in Fig.28 indicates the percentage of R^{2+} per percent octahedral occupancy necessary to maintain a neutral octahedral sheet. As the relative number of vacant positions increase, increasing the amount of negative charge, the amount of R^{3+} relative to R^{2+} must increase in order to neutralize this negative charge. The samples plotting above the line (montmorillonite) have a relative deficiency of R^{3+} ions and the layer has a net negative charge. The samples lying below the line (nontronite and saponite) have a relatively high R^{3+} content and the octahedral sheet has a net positive charge.

Attapulgite and sepiolite data plot on both sides of the neutral line. The calculated structural formulas of these two clays are more subject to error than those of the sheet structure clays. As is suggested by their low cation exchange capacity octahedral sheets tend to approach neutrality. The amount of R^{3+} required to maintain neutrality is less if OH^- substitues for O^{2-}.

The data for sepiolite are calculated for both a nine-position octahedral sheet (Nagy and Bradley, 1955) and an eight-position sheet (Brauner and Preisinger, 1956). The latter data conform to the trend established by the data for the other clays.

There are few low-temperature octahedral sheets with compositions that would plot in the center area of the compositional triangle. The low-temperature trioctahedral chlorites and perhaps Fe^{2+}-rich attapulgites contain octahedral sheets with compositions of this type.

Octahedral-tetrahedral relations

Yoder and Eugster (1955) outlined the observed composition range of natural 2:1 clays in a triangular diagram in which the two variables were tetrahedral R^{3+} and octahedral R^{3+}. The diagram divides the various clays on the basis of the relative amounts of tetrahedral and octahedral charge. A similar plot was made using 275 analyses in order to obtain some idea of how discrete the compositional limits of the various clay groups are. It is apparent, even excluding the extreme values, that there is nearly a complete overlap of fields (Fig.29). All possible combinations of octahedral and tetrahedral charge, with the exception of two apex or extreme ratios, occur in the clay minerals. Muscovite occurs at one extreme (all tetrahedral charge), pyrophyllite at another (no charge), and celadonite (all octahedral charge), a border-line clay mineral, at the third.

A number of glauconite and celadonite samples could not be plotted on the triangle as the layer charges, calculated on the basis of 2.00 filled octahedral positions, are larger than 1.00. For samples with an octahedral population larger than 2.00, the octahedral R^{3+} values are not a true measure of the amounts of octahedral charge $(2.00-R^{3+}$ = octahedral charge). These samples would have a lower octahedral charge than indicated by the present plot; however, because some of the cations in excess of 2.00 assigned to the octahedral sheet actually belong in the interlayer position (either as ions or in a hydroxide layer), the octahedral R^{3+} value may be more indicative of the true octahedral charge than is the calculated charge.

Fig.30 shows some general relations that can be inferred from these data. A layer charge of 0.7 per $O_{10}(OH)_2$ serves to divide those clays which are predominantly expanded from those that are predominantly contracted. Another subdivision can be made on the basis of whether ferric iron or aluminum is the predominant octahedral cation. The Fe clays are concentrated in the diagonal zone cutting through the center of the triangle and flanked on either side by zones populated by Al clays. Only a few

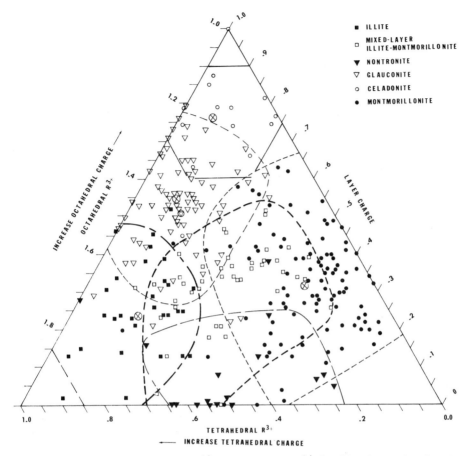

Fig.29. Ternary plot of tetrahedral R^{3+} and octahedral R^{3+} for 275 dioctahedral 2:1 clay minerals. The layer charge is assumed to be due to tetrahedral and octahedral substitution. The octahedral sheet is assumed to have 2.00 cations per $O_{10}(OH)_2$. ⊗ = average.

of the Fe-rich clays have a tetrahedral charge larger than 0.6. There is relatively little overlap of the Fe and Al minerals on the left side of the triangle: illite, glauconite, and nontronite.

There is considerable overlap between minerals named montmorillonite and nontronite. Many of the montmorillonite samples that overlap the nontronite field are relatively high in iron but so are a number of samples that lie in the restricted montmorillonite zone. The structural formulas indicate that there is a complete gradation between montmorillonite-beidellite and nontronite so that any boundary is arbitrary.

A zone near the middle of the triangle and within the less than 0.7 charge area has been labelled "mixed". Relatively few minerals fall in this zone and this zone lies in the

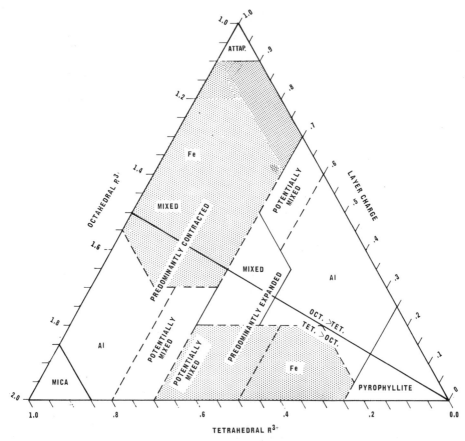

Fig. 30. Generalized classification of dioctahedral 2:1 clay minerals. Mixed refers to mixed-layering.

approximate center of the area in which the mixed-layer illite-montmorillonites lie. This is the area in which the mixed-layering is well developed. The minor component is generally greater than 30%. A few montmorillonite samples are in this area. It is quite likely that most of these samples have mixed layers and that a portion of the layers have a high enough charge to potentially contract (when potassium is available). The samples in the montmorillonite and the nontronite zones with charges greater than 0.6 and 0.5 respectively could also have some potentially contractable layers. The lower-charge boundary (0.5) for potentially mixed nontronite is determined by the larger proportion of charge originating in the tetrahedral sheet. In the contracted Al and Fe zones, those samples with the higher layer charge are more likely to be pure mineral and, as the charge decreases, the percentage of the intergrown expanded layers increases. Completely contracted clays occur principally in the Al zone (illite) illustrating the more effective strength of the tetrahedral charge compared to the octahedral charge. Glauconites always contain some expanded layers.

Mixed-layer refers to intergrowths of layers that have a sufficiently high charge to contract to approximately 10 Å and those which contain an interlayer of water (and usually a lower charge). This is gross, obvious interlayering. Even in those clays that have all layers contracted or all expanded, it is extremely likely that most are composed of two or more phases or types of layers.

The general vagueness of the various boundaries is due, in part, to the prevalence of mixed-layer clays. Contractable layers must have a charge range from 0.6 to 1.0 and expandable layers from 0.25 to 0.6 or 0.7. Most clays are a mixture of these two types of layers and a charge value of 0.7 for a clay may indicate that all layers have this charge or, more likely, that it represents a mixture of layers having values larger and smaller than 0.7.

Most of the celadonite samples lie in the area where some mixed-layering is to be expected. Although celadonite is commonly considered to be non-mixed, the literature suggests that little effort has been made to establish this. Of the 15 analyses examined by Wise and Eugster (1964) six reported adsorbed water and in the others it was not determined. In any event, the sheet structure of the celadonite is distinctly different from that of the other 2:1 dioctahedral clays (Radoslovich,1963a). It has a very thick octahedral sheet; all three octahedral positions are of equal size (in the other 2:1 dioctahedral clay; the two filled positions are smaller than the vacant position); and the interlayer separation is larger than in other contracted 2:1 dioctahedral micas (Radoslovich,1963a).

In Fig.30, the dividing line between predominantly octahedral charge and predominantly tetrahedral charge approximately coincides with the boundaries separating the Al and Fe clays (illite-glauconite and montmorillonite-nontronite). Actually a simplified division (Fig.31) based on the 0.7 charge boundary and the boundary between predominantly octahedral and predominantly tetrahedral charge coincides well with the divisions based on the plotted data.

These diagrams indicate that when the total layer charge is less than 0.7, Al will be the dominant cation when the seat of the charge is largely in the octahedral sheet; as the predominant charge shifts to the tetrahedral sheet the larger Fe ion substitutes for Al in the octahedral sheet. Radoslovich (1962) found that montmorillonite was the only layer silicate in which tetrahedral Al caused the layer to increase in size in the "b" direction. He explains this by suggesting:

"...suppose that in all the layer silicates, the tetrahedral layers exert a very small expansive force (when $\alpha > 0$). In kaolins, there is only one tetrahedral layer per octahedral layer, and in micas the interlayer cation dominates the tetrahedral twist. But in montmorillonites the small force due to *two* tetrahedral layers per octahedral layer must just have a noticeable effect."

Thus, it appears that for the low-charged dioctahedral 2:1 clays, as tetrahedral Al increases and the tetrahedral sheets expand in the "b" direction there is a tendency for this expansion to be matched by substitution of the large Fe^{3+} ion in the octahedral

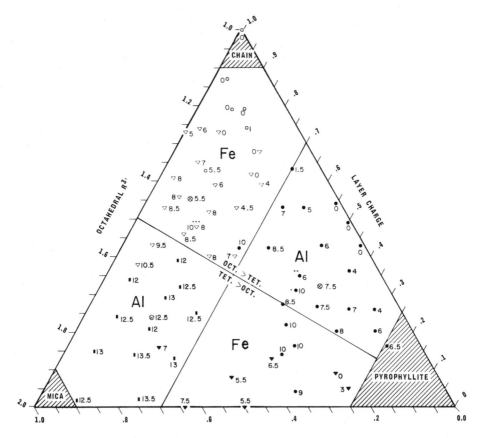

Fig.31. Idealized subdivision of the dioctahedral 2:1 clay minerals Numbers refer to amount of calculated tetrahedral twist. * = no octahedral Fe; ** = high octahedral Fe^{3+}; *** = high octahedral Al.

sheet. That this latter substitution is effective in adjusting the difference in size between the two types of sheets is suggested by the relatively small tetrahedral twist in nontronites.

This attempt at size adjustment between the two types of sheets is illustrated by the fact that very few Fe-rich 2:1 clays contain less than 0.2 tetrahedral R^{3+}. When increased Al in the tetrahedral sheets is compensated by substitution of Mg in the octahedral sheet, the octahedral charge increases and the overall layer charge is increased. This allows strong K-bonds to be developed and exert an influence on layer dimensions. Once the layer charge is larger than 0.7 and predominantly tetrahedral in origin, the Al dominated octahedral sheet represents the stable phase; such a clay approaches muscovite in composition and charge distribution and its stability is determined by the same complex balance of interlocking strong bonds ascribed to muscovite by Radoslovich (1963a).

When the tetrahedral charge is larger than 0.6, the octahedral sheet cannot adjust to the increased size of the tetrahedral sheet merely by increased size of the large Fe^{3+} ion. In this situation, it is not only necessary to substitute larger cations for octahedral Al but large divalent cations (Mg^{2+}) so that an octahedral charge is created. The resultant layer charge allows interlayer potassium to aid in adjusting the size of the two types of sheets. When the tetrahedral charge is large enough (muscovite), the octahedral charge is not required.

As the predominant charge shifts from the tetrahedral to the octahedral sheet (illite to glauconite and celadonite), anion-anion repulsion is increased (octahedral sheet of celadonite 2.48 Å thick as compared to 2.21 Å for muscovite) and the larger Fe^{3+} ion in octahedral coordination represents the more stable phase.

As the relative proportion of large cations (Fe^{3+}, Fe^{2+} Mg) in the octahedral sheet increases and the layer charge is large enough to cause the layers to be predominantly contracted, some Al substitution in the tetrahedral sheet is necessary to maintain the size difference between the sheets and allow for tetrahedral twisting. Radoslovich and Norrish (1962) have calculated that a celadonite with no tetrahedral substitution would have no tetrahedral twist and the K^+ would penetrate the oxygen sheets to such an extent that there would be "an impossibly close approach of successive layers". When tetrahedral Al values are larger than approximately 0.5–0.6 for those clays with a high octahedral charge, the amount of tetrahedral twist may become excessive. To accommodate additional tetrahedral Al it is necessary to decrease the octahedral charge so the octahedral sheet can expand in the "b" direction relatively more than the "c" direction.

When the tetrahedral charge is low and the R^{3+}/R^{2+} ratio approaches unity (upper apex of triangle), the only way to adjust the size of the octahedral sheet so that the tetrahedra can be induced to twist is to increase the amount of Al at the expense of Fe^{3+}. However, Radoslovich (1963a) states: "If Al is substituted for Fe^{3+} then the average cation–oxygen bonds are correspondingly shortened, and the octahedral cations brought closer together – in fact, unduly close".

It is suggested that under these conditions the strain in the 2:1 layer is such that it is relieved by the inversion of the tetrahedra and the formation of a chain structure (attapulgite-type). Part of the octahedral charge may be satisfied by OH⁻ proxying for oxygen. That this may happen is suggested by the fact that in attapulgite the observed hydroxyl water is larger than the calculated hydroxyl water (4–7% versus 2.15%: Caillère and Hénin, 1961a). It is possible that some of the Al-montmorillonites with a high layer charge and a high octahedral charge may be a mixture of chain and sheet structures of varying b-axis widths. Bradley (1955) has suggested that such structural irregularities, which he termed faults, are common in the trioctahedral chlorites.

To obtain some additional support for the partitioning suggested in Fig.31 an attempt was made to determine some relative tetrahedral twist values. Radoslovich and Norrish (1962) showed that the ratio:

$(Fe^{2+} + 0.853\ Fe^{3+} + 0.455\ Mg + 0.43\ Ti)/Al$ tetrahedral

is a good measure of the ratio $b_{oct.}/b_{tet.}$. They showed a good linear relation existed between this value and the calculated layer separation, which is largely controlled by the amount of tetrahedral rotation. Using their data for the dioctahedral clays and micas, a plot was made of calculated rotation versus calculated $b_{oct.}/b_{tet.}$ using the above formula (Fig.32). Although the absolute values may have appreciable error, the relative differences appear to be real.

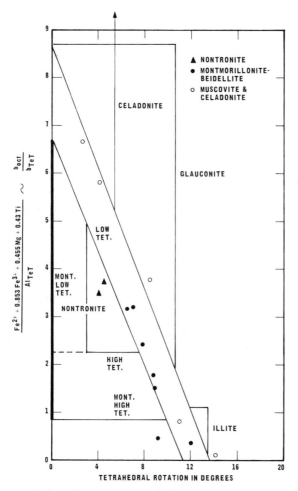

Fig.32. Plot of calculated degree of tetrahedral rotation versus calculated b oct./b tet. values for the various clay minerals. Linear relations are based on data from Radoslovich and Norrish (1962). Triangles show range of calculated tetrahedral rotation for the various clay groups (see Fig.31).

The data for the expanded and contracted minerals plot as two separate linear relations with contracted clays having larger tetrahedral rotation values for given $b_{oct.}/b_{tet.}$ values than the expanded clays. This is presumably due to the K which aids the tetrahedral rotation in the contracted clays.

Using the structural formulas that were used to plot Fig.29, the $b_{oct.}/b_{tet.}$ ratio was calculated for a number of clays and the amount of tetrahedral rotation estimated from the graphs in Fig.32. Some of these values are shown in Fig.31. The montmorillonites with a low tetrahedral Al and low octahedral R^{3+} have high ratio values and presumably a low degree of tetrahedral rotation ($0°-1.5°$). As the amount of octahedral R^{3+} increases, Mg decreases, the octahedral sheet becomes smaller, and the amount of tetrahedral rotation increases ($6.5°$).

The amount of rotation systematically increases (attaining a maximum value of approximately $10°$) concomitantly with an increase in the amount of tetrahedral Al and an increase in size by the tetrahedral sheet. When much of the octahedral Al is replaced by the larger Fe^{3+} (nontronite), the amount of rotation decreases. As the amount of tetrahedral Al increases, the amount of octahedral R^{3+} remains relatively constant and the tetrahedral rotation increases from $0°-3°$ to $7.5°$.

When there are more than 0.5–0.6 tetrahedral Al, the layer charge becomes strong enough to cause K-fixation and contraction (illite) and the amount of tetrahedral rotation is nearly doubled ($13°$). Along with the increase in tetrahedral Al there is an increase in the amount of octahedral Al and Mg at the expense of Fe^{3+}. The analyses suggest that this is a continuous series and much of the abrupt increase in rotation is due to the increased layer charge and the fixation of the K ion. There appears to be no good reason why Fe^{3+} illites could not exist. An illite-type clay with an octahedral sheet composed predominantly of Fe^{3+} ($Fe^{3+}_{1.80}$ $Al_{0.10}Mg_{0.10}$ and $Fe^{3+}_{1.80}Mg_{0.20}$ with 0.6 tetrahedral Al) would have a tetrahedral rotation of approximately $11°-9°$, which is larger than that for most glauconites.

Present data indicate that Fe^{3+}-rich low-charge clays increase their layer charge by increasing the Mg and Fe^{2+} content of the octahedral sheet at the expense of Fe^{3+} more so than of Al. The average Al content of glauconite and celadonite is similar to that of nontronite, but the Fe^{3+} values are lower. With increased octahedral charge there is an increase anion-anion repulsion and the octahedral sheet increases relatively more in the "c" direction than the "b" direction, which also favors the large cations. Thus, relatively less tetrahedral Al is required to afford the sheet size differential to allow sufficient tetrahedral rotation to lock the K into place.

The calculated tetrahedral rotation for glauconites with high octahedral R^{3+} values ranges from 8 to $10°$. There appears to be no overlap of the illite values ($12°-13.5°$). As the amount of octahedral R^{3+} decreases, the octahedral sheet increases in size and charge and rotation values decrease. As the amount of tetrahedral Al decreases, the sheets become similar in size and the amount of rotation approaches zero (K cannot be locked in position to provide sufficient layer separation). As octahedral R^{3+} decreases and octahedral charge increases the amount of tetrahedral Al would have to increase to afford the size differential for tetrahedral twist. This would produce a layer charge larger than one, which is presumably an unstable condition for low-temperature layer structures. Further changes would require material changes in the makeup of the octahedral and/or tetrahedral sheets.

Layer structure clays have four tetrahedra per three octahedral positions. When the limits suggested above are reached, there is little additional adjustment that can be accomplished by compositional changes. It is necessary for the tetrahedron/octahedron ratio to change. This is accomplished by increasing the tetrahedron/octahedron ratio to 8/5 (attapulgite). This structural change is apparently an over-compensation and to bring the sheets into closer accord, the amount of tetrahedral Al is decreased. Although the amount of octahedral R^{2+} is increased the octahedral charge is decreased (favoring stretching in the "b" direction) by the addition of H^+. However, the size differential caused by the 8/5 ratio is apparently too large to be adjusted by compositional changes and the strain build-up over an interval of five octahedral positions is sufficient to cause tetrahedral inversion.

When the relative amount of large cations (Fe^{3+}, Fe^{2+}, Mg) in the octahedral sheet is increased (80 to 100% of the octahedral cations), the layer strain is slightly reduced and the octahedral sheet can extend to 8 or 9 positions before forcing the tetrahedral inversion (sepiolite). It seems likely that the octahedral sheets of attapulgite and sepiolite have a much wider compositional range than indicated by the present limited analyses and that they occupy the intermediate position between the dioctahedral and trioctahedral clays.

Chain structure

Attapulgites have less octahedrally coordinated alumina than is found in the montmorillonites. The magnesium content of the octahedral sheet is 2 to 4 times as abundant as in the montmorillonites. The iron contents are similar. The average Al_2O_3/MgO ratio for illites is 9.6, and for montmorillonites 6.7. Attapulgite values range from 2.5 to 0.48. The average ratio octahedral Al/octahedral Mg are respectively 5.4, 4.3, and 1.8 to 0.4.

Radoslovich (1963a) found that the 2M muscovite structure required that a minimum of 1.7 of the octahedral sites be filled with Al. The lower limit of 1.7 Al is equivalent to 85% of the two symmetrically related or occupied sites being filled in a stable muscovite structure. For 29 illites, an average of 1.55 Al per three octahedral sites was found. This is 77% occupancy of the two occupied sites. Total trivalent ions (Al + Fe^{3+}) averaged 1.77 (88% occupancy). The average Al per three sites for 101 montmorillonites is 1.49 (74.5%) and for Al and Fe^{3+}, 1.68 (84%). Frequency-distribution graphs of data from these two minerals groups indicate that the large majority contain between 1.3 and 1.7 Al per three sites. There are relatively abrupt decreases in frequency at these values (Fig.2,14) suggesting that they are significant boundary values. The minimum value of 1.3 is equivalent to 65% occupancy of the two occupied sites as compared to a minimum of 85% for 2M muscovite. The size of the cation is considered to be more important in controlling the size of the octahedral sheet than the charge of the cation. Thus, it is the amount of octahedral Al rather than Al + Fe^{3+} that most closely reflects the amount of strain in the octahedral sheet.

In attapulgite, only four of the proposed five octahedral sites are occupied. The seven structural formulas caculated by Caillère and Hénin (1961) indicate that Al fills from 1.13 to 2.34 of these five sites or between 28 and 59% of the four occupied sites. (Al + Fe^{3+} values range from 31% to 62%.) The amount of tetrahedral Al per four positions ranges from 0.05 to 0.34 and averages 0.165. When four of the formulas were recalculated on the basis of the measured water loss in different temperature ranges, the average tetrahedral Al was 0.085 with two of the four showing no tetrahedral substitution. The amount of octahedral Al per four positions averaged 50.1%. On the basis of these admittedly few data, it would appear that the chain-structure attapulgites are the most likely clay minerals to have an Al/Mg ratio near 1. They also have relatively little tetrahedral substitution.

In the dioctahedral 2:1 sheet-structure silicate with the occupied sites more than 85% occupied by Al, the structure seems to be able to compensate for the internal strain and can grow to a considerable size. The Al octahedral occupancy values of muscovite (>1.7) and the 2:1 dioctahedral clays (1.3–1.7) indicate that there is little overlap. It is likely that the decreased amount of tetrahedral twist induced by increasing the size of the octahedral cations and octahedral charge (decreasing Al) determines that a clay-size rather than a larger mineral will form. The R^{3+} occupancy value can be less than 1.3 when the larger Fe^{3+} is substituted for Al. When Al occupancy values are less than 1.3 (65%), in the absence of appreciable iron, the internal strain is such that growth is in only one direction. The width of the layer is restricted to five octahedral sites. Sufficient layer strain accumulates within this five-site interval such that the silica tetrahedral sheet is forced to invert to accommodate the strain.

REFERENCES

Abdul-Latif, N. and Weaver, C.E., 1969. Kinetics of acid-dissolution of palygorskite (attapulgite) and sepiolite. *Clays Clay Miner.*, 17:169–178.

Adler, H.H. and Kerr, P.E., 1950. Infrared spectra of reference clay minerals. *Am. Pet. Inst. Proj. 49, Clay Miner. Stand., Prelim. Rep.*, 8.

Ahrens, L.H., 1954. The lognormal distribution of the elements. *Geochim. Cosmochim. Acta*, 5:49–73.

Alexander, L.T., Faust, G.T., Hendricks, S.B., Insley, H. and McMurdie, H.F., 1943. Relationship of the clay minerals halloysite and endellite, *Am. Mineralogist*, 28:1–18.

Alietti, A., 1956. Il mineral a strati misti saponite-talco di Monte Chiaro (Val di Taro, Appenneno Emiliano). *Accad. Naz. Lincei, Rend. Cl. Sci. Fis. Math. Nat. Ser. 8*, 21 (3/4):201–207.

Alietti, A., 1957. Il minerale interlaminato chlorite-saponite di Gotra. *Accad. Sci. Lett. Arti Modena*, 15:3–14.

Alietti, A., 1960. Sudi una nontronite poco ferrifera di Chiampo (Lesseni). *Soc. Tipogr. Editr. Modenese*, 1960:33–44.

Alietti, A. and Alietti, L., 1962. Su due diversi tipi di montmorillonite riconosciuti nella bentonite di Gemmano (Forli). *Period. Mineral.*, 31(2/3):261–286.

Al Raivi, A.H., Jackson, M.L. and Hole, F.D., 1969. Mineralogy of some arid and semiarid soils of Iraq. *Soil Sci.*, 107:480–486.

Altaie, F.H., Sys, C. and Stoops, G, 1969. Soil groups of Iraq: their classification and characterization. *Pedologie*, 9:65–148.

Altschuler, Z.S., Dwornik, E.J. and Kramer, H., 1963. Transformation of montmorillonite to kaolinite during weathering. *Science*, 141:148–152.

Ames, L.L. and Goldich, S.S., 1958. A contribution on the Hector, California, bentonite deposit. *Econ. Geol.*, 53:22.

Andreatta, C., 1949. Studio di un intersante giacimento di riempimento di argille montmorillonite idrotermali (Vallortigara-Posina, Schio). *Mem. Accad. Sci. Inst. Bologna, It. Sci. Fiz., Sez. Sci. Nat., Ser. 10*, 5:M.A. 11–547.

Aomine, S. and Jackson, M.L., 1959. Allophane determination in Avdo soils by cation-exchange capacity delta values. *Proc. Soil. Sci. Soc. Am.*, 23:210–214.

Arrhenius, G., 1963. Pelagic sediments. In: M.N. Hill (Editor), *The Sea*. Interscience, New York, N.Y., pp. 655–727.

Bailey, R.J. and Atherton, M.P., 1969. The petrology of a glauconitic sandy chalk. *J. Sediment. Petrol.*, 39:1420–1431.

Bailey, S.W. and Langston, R.B., 1969. Anauxite and kaolinite structures identical. *Clays Clay Miner.*, 17:241–243.

Bailey, S.W. and Tyler, S.A., 1960. Clay minerals associated with the Lake Superior iron ores. *Econ. Geol.*, 55:150–175.

Bailey, S.W., Brindley, G.W., Johns, W.D., Martin, R.T. and Ross, M., 1971. Report of Clay Minerals Society Nomenclature Committee. *Clays Clay Miner.*, 19:132–134.

Ball, D.F., 1966. Chlorite clay minerals in Ordovician pumice-tuff and derived soils in Snowdonia, north Wales. *Clay Miner.*, 6:195–209.

Bannister, F.A., 1943. Brammalite (sodium-illite), a new mineral from Llandebie, south Wales. *Mineral. Mag.*, 26:304–307.

Bannister, F.A. and Whittards, W.F., 1945. A magnesium chamosite from the Wenlock Limestone of Wickwar, Gloucestershire. *Mineral. Mag.*, 27:101–110.
Bardossy, Gy, 1959. The geochemistry of Hungarian bauxites, 1, 2, 3, 4. *Acta Geol. Acad. Sci. Hung.*, 5:103–155; 255–285; 6:1–53.
Barnhisel, R.I. and Rich, C.I., 1963. Gibbsite formation from aluminum-interlayers in montmorillonite. *Soil Sci. Soc. Am., Proc.*, 27:632–635.
Barshad, I. and Kishk, F.M. 1969. Chemical composition of soil vermiculite clays as related to their genesis. *Contrib. Mineral. Petrol.*, 24:136–155.
Barshad, I., Halevy, E., Gold, H.A. and Hagin, J., 1956. Clay minerals in some limestone soils from Israel. *Soil Sci.*, 81:423–437.
Bassett, W.A., 1959. Origin of the vermiculite deposite at Libby, Montana. *Am. Mineralogist*, 44:282–299.
Bassett, W.A., 1963. The geology of vermiculite occurrences. *Clays Clay Miner., Proc.*, 10:61–69.
Bates, T.F., 1947. Investigation of the micaceous minerals in slate. *Am. Mineralogist*, 32:625–636.
Bates, T.F., 1959. Morphology and crystal chemistry of 1:1 layer lattice silicates. *Am. Mineralogist*, 44:78–114.
Bates, T.F., 1962. Halloysite and gibbsite formation in Hawaii. *Clays Clay Miner. Proc.*, 9:315–328.
Bates, T.F., Hildebrand, F.A. and Swineford, A., 1949. Electron microscopy of clay minerals (abst.): *Am. Mineralogist*, 35:463–484.
Bates, T.F., Hildebrand, F.A. and Swineford, A., 1950. Morphology and structure of endellite and halloysite. *Am. Mineralogist*, 35:463–484.
Bayramgil, O., Hügi, Th. and Nowacki, W., 1952. Uber ein Seladonitvorkommen im Gebiete von Zonguldak (Turkei). *Schweiz. Mineral. Petrog. Mitt.*, 32:243–250.
Belyankin, D.S. and Ivanova, V.P., 1936. Conversion of kaolin when heated. In:*Symposium to V.I. Vernadskii*, 1:535–562.
Bentor, Y.K. and Kastner, M., 1965. Notes on the mineralogy and origin of glauconite. *J. Sediment. Petrol.*, 35:155–166.
Birrell, K.S. and Gradwell, M., 1956. Ion-exchange phenomena in some soils containing amorphous mineral constituents. *J. Soil. Sci.*, 7:130–147.
Blokh, A.M. and Siderenko, G.A., 1960. Nefedyevite from the Transbaikal. *Dokl. Akad. Nauk S.S.S.R.*, 135(3):701–704.
Bonatti, E. and Joensuu, O., 1968. Palygorskite from Atlantic deep sea sediments. *Am. Mineralogist*, 53:975–983.
Bonatti, S. and Gallitelli, P., 1950. Metahalloysite nelle farine fossili di Bagnoregio (Viterbo). *Atti Soc. Toscana Sci. Nat., Ser. A*, 57:10 pp.
Borchert, H. and Braun, H., 1963. Zum Chemismus von drei Glauconittypen. *Chem.Erde*, 23:82–90.
Bradley, W.F., 1940. Structure of attapulgite. *Am. Mineralogist*, 25:405–410.
Bradley, W.F., 1945. Diagnostic criteria for the recognition of clay minerals. *Am. Mineralogist*, 30:704–713.
Bradley, W.F., 1950. The alternating layer sequence of rectorite. *Am. Mineralogist*, 35:590–595.
Bradley, W.F., 1955. Structural irregularities in hydrous magnesium silicates. *Proc. Natl. Conf. Clays Clay Miner., 3rd. – Natl. Acad. Sci. Natl. Res. Counc., Publ.*, 395:94–102.
Bradley, W.F. and Fahey, J.J., 1962. Occurrence of stevensite in the Green River Formation of Wyoming. *Am. Mineralogist*, 47:996–998.
Bradley, W.F. and Krause, J.T., 1957. Structure in ionic solutions, 1. *J. Chem. Phys.*, 27:304–308.
Bradley, W.F. and Weaver, C.E., 1956. A regularly interstratified chlorite-vermiculite clay mineral. *Am. Mineralogist*, 41:497–504.
Brammall, A., Leech, J.G.C. and Bannister, F.A., 1937. The paragenesis of cookeite and hydromuscovite associated with gold at Ogofau, Carmarthenshire. *Mineral. Mag.*, 24:507–519.
Brauner, K. and Preisinger, A., 1956. Structure of sepiolite. *Mineral. Petrogr. Mitt.*, 6:120–140.
Brindley, G.W., 1951. The crystal structure of some chamosite minerals. *Mineral. Mag.*, 29:502–525.

Brindley, G.W., 1955. Stevensite — a mixed-layer mineral. *Am. Mineralogist*, 40:239–247.
Brindley, G.W., 1956. Allevardite, a swelling double layer mica mineral. *Am. Mineralogist*, 41:91–103.
Brindley, G.W., 1961a. Chlorite minerals. In: G. Brown (Editor), *The X-ray Identification and Crystal Structures of Clay Minerals*. London Mineral. Soc., London, pp. 242–296.
Brindley, G.W., 1961b. Kaolin, serpentine and kindred minerals. In: G. Brown (Editor), *The X-Ray Identification and Crystal Structures of Clay Minerals*. Londen Mineral. Soc., London, pp. 51–131.
Brindley, G.W. and Gillery, F.H., 1954. A mixed-layer kaolin-chlorite structure. *Proc. Natl. Conf. Clays Clay Miner., 2nd — Natl. Acad. Sci. Natl. Res. Counc., Publ.*, 327:349–353.
Brindley, G.W. and Gillery, F.H., 1956. X-ray identification of chlorite species. *Am. Mineralogist*, 41:169–186.
Brindley, G.W. and Udagawa, S., 1960. High-temperature reactions of clay mineral mixtures and their ceramic properties, 1. Kaolinite-mica-quartz mixtures with 25 weight % quartz. *J. Am. Ceram. Soc.*, 43:59–65.
Brindley, G.W. and Youell, R.F., 1953. Ferrous chamosite and ferric chamosite. *Mineral. Mag.*, 30:57–70.
Brindley, G.W., Oughton, B.M. and Robinson, K., 1950. Polymorphism of the chlorites, 1. Ordered structures. *Acta Cryst.*, 3:408–416.
Brown, B.E. and Bailey, S.W., 1962. Chlorite polytypism, 1. Regular and semi-random one-layer structures. *Am. Mineralogist*, 47:819–850.
Brown, B.E. and Bailey, S.W., 1963. Chlorite polytypism, 2. Crystal structure of a one-layer Cr-chlorite. *Am. Mineralogist*, 48:42–61.
Brown, G., 1953. The dioctahedral analoque of vermiculite. *Clay Miner.*, 2:64–70.
Brown, G. and Weir, A.H., 1963. The identity of rectorite and allevardite. *Int. Clay Conf., Proc., 1963*, 1:27–35.
Brydon, J.E., Clark, J.S. and Osborne, V., 1961. Dioctahedral chlorite. *Can. Mineralogist*, 6:595–609.
Bundy, W.M., Johns, W.D. and Murray, H.H., 1965. Interrelationships of physical and chemical properties of kaolinites (abstr.). *Clay Miner. Soc., 2nd meeting*, p.34.
Burst, J.F., 1958. Mineral heterogeneity in "glauconite" pellets. *Am. Mineralogist*, 43:481–497.
Burst, J.F., 1959. Postdiagenic clay mineral environmental relationships in the Gulf Coast Eocene. *Proc. Natl. Conf. Clays Clay Miner., 6th — Natl. Acad. Sci. Natl. Res. Counc., 1957*: 327–341.
Bystrom, A.M., 1956. Mineralogy of the Ordovician bentonite beds at Kinnekulle, Sweden. *Sver. Geol. Undersökn. Arsbok.*, 48:1–62.
Cahoon, H.P., 1954. Saponite near Milford, Utah. *Am. Mineralogist*, 39:222–230.
Caillère, S., 1934. Observations sur la composition chimique des palygorskites. *Compt. Rend.* 198:1795–1798.
Caillère, S., 1936a. Etude de quelques silicates magnésiens a facies asbestiform ou papyrace n'appartenant pas au groupe de l'entigorite. *Bull. Soc. Franc. Mineral.*, 59:353–374.
Caillère, S., 1936b. Thermal studies. *Bull. Soc. Franc. Mineral.*, 59:163–326.
Caillère, S., 1951. Palygorskite from Tafraout. *Compt. Rend.*, 233:697–698.
Caillère, S., 1953. Synthesis à basse température de phyllites ferrifères. *Compt. Rend.*, 237:1424–1426.
Caillère, S., 1954. Synthesis de quelques phyllites nickelifères. *Compt. Rend.*, 239: 1535–1536.
Caillère, S., 1962. A new type of chlorite. *Bull. Groupe Franc. Argils.*, 13:35–37.
Caillère, S. and Hénin, S., 1949. Experimental formation of chlorites from montmorillonites. *Mineral. Mag.*, 28:612–620.
Caillère, S. and Hénin, S., 1950. Quelques remarques sur la synthèse des mineeraux argileux. *Congr. Int. Sci. Sol, 4me*, 1:96–98.
Caillère, S. and Hénin, S., 1951. The properties and identification of saponite (Bowlingist). *Clay Miner.*, 1:138–144.
Caillère, S. and Hénin, S., 1961a. Palygorskite. In: G. Brown (Editor), *The X-Ray Identification and Crystal Structures of Clay Minerals*. London, London Mineral. Soc., London, pp. 343–353.

REFERENCES

Caillère, S. and Hénin, S., 1961b. Sepiolite. In: G. Brown (Editor), *The X-Ray Identification and Crystal Structures of Clay Minerals*. London Mineral. Soc., London, pp. 325–342.
Caillère, S., Hénin, S. and Esquevin, J., 1953. Récherches sur la synthesis des mineraux argileux. *Bull. Soc. Franc. Mineral. Crist.*, 76:300.
Carr, K., Grimshaw, R.W. and Roberts, A.L., 1953. Hydrous mica from Yorkshire Fireclay: *Mineral. Mag.*, 30:139–144.
Carroll, D. and Starkey, H.C., 1960. Effect of seawater on clay minerals. *Proc. Natl. Conf. Clays Clay Minerals, 7th–Natl. Acad. Sci. Natl. Res. Counc., 1958:* 80–101.
Carstea, D.D., 1968. Formation of hydroxy-Al and -Fe interlayers in montmorillonite and vermiculite: influence of particle size and temperature. *Clays Clay Miner.*, 16:231–238.
Carstea, D.D., Harward, M.E. and Knox, E.G., 1970. Formation and stability of hydroxy-Mg interlayers in phyllosilicates. *Clays Clay Miner.*, 18:213–222.
Christ, C.L., Hathaway, J.C., Hostetler, P.B. and Shephard, A.O., 1969. Palygorskite: new X-ray data. *Am. Mineralogist*, 54:198–205.
Chukhrov, F.V. and Anosov, F. Ya., 1950. Medmontite, a copper-bearing montmorillonite mineral. *Zapiski Vses. Mineral. Obshch.*, 79:23–27.
Chukhrov, F.V., Berkhin, S.I., Ermilova, L.P., Moelva, V.A. and Rudnitskaya, E.S., 1963. Allophanes from some deposits of the U.S.S.R. *Int. Clay Conf., Stockholm*, 1963: 19–28.
Chukhrov, F.V., Zvyagin, B.B., Gorshkov, A.I. and Yermilova, L.P., 1970. Copper-bearing halloysites. *Int. Geol. Rev.*, 12:34–38.
Cimbálniková, A., 1971. Chemical variability and structural heterogeneity of glauconites. *Am. Mineralogist*, 56:1385–1392.
Cole, W.F., 1966. A study of a long-spacing mica-like mineral. *Clay Miner.*, 6:261–281.
Collet, L.W. and Lee, G.W., 1906. Recherches sur la glauconie. *Proc. R. Soc. Edinburgh*, 26:238–278.
Conway, E.J., 1942. Mean geological data in relation to oceanic evolution. *Proc. R. Irish. Acad., Sect. B.*, 48:119–159.
Curtis, C.D., Brown, P.E. and Somogyi, V.A., 1969. A naturally occurring sodium vermiculite from Unst, Shetland. *Clay Miner.*, 8:15–19.
Dana, J.W., 1892. *The System of Mineralogy*. Wiley, New York, N.Y., 618 pp.
Deer, W.A., Howie, R.A. and Zussman, J., 1962. *Rock Forming Minerals, 3. Sheet Silicates*. Wiley, New York, N.Y., 270 pp.
DeKimpe, C.R., 1969. Crystallization of kaolinite at low temperature from an alumino- silicic gel. *Clays Clay Miner.*, 17:37–38.
DeKimpe, C.R. and Fripiat, J.J., 1968. Kaolinite crystallization of H-exchanged zeolites: *Am. Mineralogist*, 53:216–230.
DeKimpe, C.R., Gastuche, M.C. and Brindley, G.W., 1961. Ionic coordination in alumino- silicic gels in relation to clay mineral formation. *Am. Mineralogist*, 46:1370–1381.
DeKimpe, C.R., Gastuche, M.C. and Brindley, G.W., 1964. Low-temperature synthesis of kaolin minerals. *Am. Mineralogist*, 49:1–16.
De Lapparent, J., 1935. Sur un constituant essential des terres à foulon. *Compt. Rend.*, 201:481–483.
De Mumbrum, L.E. and Chesters, G., 1964. Isolation and characterization of some soil allophanes. *Proc. Soil Sci. Soc. Am.*, 28:355–359.
DeSousa Santos, P., Brindley, G.W. and De Sousa Santos, H., 1965. Mineralogical studies of kaolinite-halloysite clays, 3. A fibrous kaolin mineral from Piedade, Sao Paulo, Brasil. *Am. Mineralogist*, 50:619–628.
De Sousa Santos, P., De Sousa Santos, H. and Brindley, G.W., 1966. Mineralogical studies of kaolinite-halloysite clays, 4. A platy mineral with structural swelling and shrinking characteristics. *Am. Mineralogist*, 51:1640–1648.
Deudon, M., 1955. La chamosite orthorhombique du minerai de Sante-Barbe, Conche Grise. *Bull. Soc. Franc. Mineral. Crist.*, 78:474–480.
Doelter, C., 1914. *Doelter's Handbuch der Mineral-chemie, 2*.

Dolcater, D.L., Syers, J.K. and Jackson, M.L., 1970. Titanium as free oxide and substituted forms in kaolinite and other soil minerals. *Clays Clay Miner.*, 18:71–79.
Drennan, J.A., 1963. An unusual occurrence of chamosite. *Clay Miner.*, 5:382–391.
Droste, J.B., 1956. Alteration of clay minerals by weathering in Wisconsin tills. *Geol. Soc. Am., Bull.*, 67:911–918.
Dunham, K.C., Claringbull, G.F. and Bannister, F.A., 1948. Dickite and collophane in the magnesium limestone of Durham. *Mineral. Mag.*, 28:338–342.
Dyadchenko, M.G. and Khatuntzeva, A. Ya., 1955. The genesis of glauconite. *Dokl. Akad. Nauk S.S.S.R.*, 101 (1):151–153.
D'yakonov, Yu.S. and L'vova, I.A., 1967. Transformation of trioctahedral micas into vermiculite. *Dokl. Akad. Nauk S.S.S.R.*, 175:127–129. (Translated from O. Prevrashchenii trioktaedricheskikh slyud v vermikulit. *Dokl. Akad. Nauk S.S.S.R.*, 175:432–434.)
Earley, J.W., Brindley, G.W., McVeagh, W.V. and Van den Heuvel, R.C., 1956. A regularly interstratified montmorillonite-chlorite. *Am. Mineralogist*, 41:258–267.
Eberl, D., 1970. Low-temperature synthesis of kaolinite from amorphous material at neutral pH. *Conf. Clay Minerals Soc., 19th, Abstr.*, p. 17.
Efremov, N.E., 1936. Ferri-halloysite from the ore beak of the Taman Peninsula. *Bull. Soc. Naturalistes Moscow, Sect. Geol.*, 14(3):277–282.
Eggleston, R.A. and Bailey, S.W., 1967. Structural aspects of dioctahedral chlorite. *Am. Mineralogist*, 52:673–689.
Ehlmann, A.J., Hulings, N.C. and Glover, E.D., 1963. Stages of glauconite formation in modern foraminiferal sediments. *J. Sediment. Petrol.*, 33:87–96.
Elgabaly, M.M., 1962. The presence of attapulgite in some soils of the western desert of Egypt. *Soil Sci.*, 93:387–390.
Emkew, M.P., 1958. Sauconite from Altyn-Tojkan. *Zap. Uzbek Otdel Vses. Mineral. Obshch.*, 12:79–84; *Chem. Abstr.*, 53:16839.
Evernden, J.F., Curtis, G.H., Obradovich, J. and Kistler, R., 1961. On the evaluation of glauconite and illite for dating sedimentary rocks by the potassium-argon method. *Geochim. Cosmochim. Acta*, 23:78–99.
Fahey, J.J., Ross, M. and Axelrod, J.M., 1960. Loughlinite, a new hydrous sodium magnesium silicate. *Am. Mineralogist*, 45:270–281.
Faust, G.T., 1951. Thermal analysis and X-ray studies of sauconite and of some zinc minerals of the same paragenetic association. *Am. Mineralogist*, 36:795–822.
Faust, G.T., 1955. Thermal analysis and X-ray studies of griffithite. *J. Wash. Acad. Sci.*, 45:66–70.
Faust, G.T. and Fahey, J.J., 1962. The serpentine group minerals. *U.S. Geol. Surv., Prof. Pap.*, 384-A: 92 pp.
Faust, G.T. and Murata, K.J., 1953. Stevensite, redefined as a member of the montmorillonite group. *Am. Mineralogist*, 38:973–987.
Faust, G.T., Hathaway, J.C. and Millot, G., 1959. Restudy of stevensite. *Am. Mineralogist*, 44:342–370.
Feitchnecht, W., 1962. Uber die Bildung von Doppelhydroxyde zwischen Zwei- und Dreiwertigen Metallen: *Helv. Chem. Acta,*, 25:555–569.
Fersmann, A., 1913. Research on magnesian silicates. *Mem. Acad. Sci. St. Petersb.*, 32:321–430.
Fieldes, M., 1956. Clay mineralogy of New Zealand soils, 2. Allophane and related mineral colloids. *N. Z. J. Sci. Tech.*, 37:336–350.
Follett, E.A.C., McHardy, W.J., Mitchell, B.D. and Smith, B.F.L., 1965. Chemical dissolution techniques in the study of soil clays, 1. *Clay Miner.*, 6:23–34.
Foster, M.D., 1951. The importance of exchangeable magnesium and cation exchange capacity in the study of montmorillonite clays. *Am. Mineralogist*, 36:717–730.
Foster, M.D., 1954. The relation between "illite", beidellite, and montmorillonite. *Proc. Natl. Conf. Clays Clay Miner., 2nd–Natl. Acad. Sci. Natl. Res. Counc., Publ.*, 327:386–397.
Foster, M.D., 1956. Correlation of dioctahedral potassium micas on the basis of their charge relations. *U.S. Geol. Surv., Bull.*, 1036-D:57–67.

Foster, M.D., 1960. Interpretation of the composition of trioctahedral micas. *U.S. Geol. Surv. Prof. Pap.*, 354-B.
Foster, M.D., 1962. Interpretation of the composition and a classification of the chlorites. *U.S. Geol. Surv., Prof. Pap.*, 414-A:27 pp.
Foster, M.D., 1963. Interpretation of the composition of vermiculites and hydrobiotites. *Clays Clay Miner., Proc.*, 10:70–89.
Foster, M.D., 1969. Studies of celadonite and glauconite. *U.S. Geol. Surv., Prof. Pap.*, 614-F:17 pp.
Fournier, R.O., 1965. Montmorillonite pseudomorphic after plagioclase in a porphyry copper deposit. *Am. Mineralogist*, 50:771–777.
Frank, H.S. and Evans, M.W., 1945. Free volume and entropy in condensed systems, 3. *J. Chem. Phys.*, 13:507–532.
Garrels, R.M. and Christ, C.L., 1965. *Solutions, Minerals and Equilibria.* Harper and Row, New York, N.Y., 450 pp.
Garrett, W.G. and Walker, G.F., 1959. The cation exchange capacity of hydrated halloysite and the formation of halloysite salt complexes. *Clay Miner.*, 4:75–80.
Gastuche, M.C., 1963. The octahedral layer. *Clays Clay Miner., Proc.*, 12:471–493.
Gaudette, H.E., 1965. Illite from Fond du Lac County, Wisconsin. *Am. Mineralogist*, 50:411–417.
Gaudette, H.E., Eades, J.L. and Grim, R.E., 1966. The nature of illite. *Clays Clay Miner. Proc.*, 13:33–48.
Gillery, F.H., 1959. The X-ray study of synthetic Mg-Al serpentines and chlorites. *Am. Mineralogist*, 44:143–152.
Gjems, O., 1963. A swelling dioctahedral clay mineral of a vermiculite-smectite type in the weathering horizons of podzols. *Clay Miner.*, 5:183–193.
Glass, J.J., 1935. The pegmatite minerals from Amelia, Virginia. *Am. Mineralogist*, 20:741–768.
Glinka, K., 1896. *Der Glauconite, sein Entstehung, sein chemischer Bestand, und die Art und Weise seiner Verwitterung.* Inst. Agrom. de Novo, Alex. Russie, St. Petersburg.
Gogishvili, V.G., 1959. Celadonite from the surroundings of Tbilisi. *Geol. Sb. Kavkaz Inst., Mineral. Syr'yal.*, 134–136.
Greene-Kelly, R., 1955. Dehydration of the montmorillonite minerals. *Mineral. Mag.*, 30:604–615.
Grim, R.E., 1953. *Clay Mineralogy.* McGraw-Hill, New York, N.Y.
Grim, R.E. and Johns, W.D., 1954. Clay mineral investigation of sediments in the northern Gulf of Mexico. *Proc. Natl. Conf. Clays Clay Miner., 2nd–Natl. Acad. Sci. Natl. Res. Counc., Publ.*, 327:81–103.
Grim, R.E. and Kulbicki, G., 1961. Montmorillonite. High-temperature reactions and classification. *Am. Mineralogist*, 46:1329–1369.
Grim, R.E., Bray, R.H. and Bradley, W.F., 1937. The mica in argillaceous sediments. *Am. Mineralogist*, 22:813–829.
Grim, R.E., Droste, J.B. and Bradley, W.F., 1961. A mixed-layer clay mineral associated with an evaporite. *Clays Clay Miner., Proc. Natl. Conf. Clays Clay Miner.*, 8(1959):228–236.
Gritzanko, G.S. and Grum-Grzhimailo, S.V., 1949. On chromium halloysite from Aidyrly deposits in southern Urals. *Mem. Soc. Russe Min., Ser. 2*, 78:61–63.
Gruner, J.W., 1935. The structural relationship of glauconite and mica. *Am. Mineralogist*, 20:699–714.
Gruner, J.W., 1936. The structure and chemical composition of greenalite. *Am. Mineralogist*, 21:449–455.
Gruner, J.W., 1944. The kaolinite structure of amesite, and additional data on chlorites: *Am. Mineralogist*, 29:422–432.
Guenot, B., 1970. Etude d'un minéral argileux du type interstratifié talc-saponite trouvé dans le Précambrien du Congo Kinshasa. *Bull. Groupe Franç. Argiles*, 22:97–104.
Hallimond, A.F., 1922. On glauconite from the greensand near Lewes, Sussex; the constitution of glauconite. *Mineral. Mag.*, 19:330–333.

REFERENCES

Hallimond, A.F., 1925. Iron ores, bedded ores of England and Wales. *Mineral. Resour. Gr. Brit., Spec. Publ.,* 29:26–27.

Hanshaw, B.B., 1964. Cation-exchange constants for clays from electrochemical measurements. *Clays Clay Miner., Proc.,* 12:397–422.

Haseman, J.F., Lehr, J.R. and Smith, J.P., 1951. Mineralogical character of some iron and aluminum phosphates containing potassium + ammonium. *Proc. Soil Sci. Soc. Am.,* 15:76–84.

Hathaway, J.C., 1955. Studies of some vermiculite-type minerals. *Proc. Natl. Conf. Clays Clay Miner., 3rd–Natl. Acad. Sci. Natl. Res. Counc., Publ.,* 395:74–86.

Hathaway, J.C. and Sachs, P.L., 1965. Sepiolite and clinoptilolite from the Mid-Atlantic Ridge. *Am. Mineralogist,* 50:852–867.

Hawkins, D.B. and Roy, R., 1963. Experimental hydrothermal studies on rock alteration and clay mineral formations. *Geochim. Cosmochim. Acta,* 27:1047–1054.

Hayes, J.B., 1970. Polytypism of chlorite in sedimentary rocks. *Clays Clay Miner.,* 18:285–306.

Heckroodt, R.O. and Roering, C., 1965. A high-alluminous chlorite-swelling chlorite regular mixed-layer clay mineral. *Clay Miner.,* 6:83–89.

Heddle, M.F., 1879. The minerals of Scotland: celadonite. *Trans. R. Soc. Edinburgh,* 29:101–104.

Hendricks, S.B., 1939a. Crystal structure of nacrite and the polymorphism of the kaolin minerals. *Z. Krist.,* 100:509–518.

Hendricks, S.B., 1939b. Random structure of layer minerals as illustrated by cronstedtite: possible iron content of kaolinite. *Am. Mineralogist,* 24:529–539.

Hendricks, S.B., 1939b. Random structure of layer minerals as illustrated by cronstedtite: possible 50:276–290.

Hendricks, S.B. and Ross, C.S., 1941. Chemical composition and genesis of glauconite and celadonite. *Am. Mineralogist,* 26:683–708.

Hénin, S., 1954. Nouveaux résultats concernant la préparation des minéraux argileux au laboratoire. Synthesis de l'antigorite. *Compt. Rend.,* 238:2554–2556.

Hénin, S., 1956. Synthesis of clay minerals at low temperature. *Proc. Natl. Conf. Clays Clay Miner., 4th–Natl. Acad. Sci. Natl. Res. Counc., Publ.,* 456:54–60.

Hénin, S. and Robichet, Q., 1953. Sur les conditions de formation des minéraux argileux par voie experimentale à basse temperature. *Compt. Rend.,* 236:517–519.

Hénin, S., Esquevin, J. and Caillère, S., 1954. Sur la fabrosite de certains minéraux de nature montmorillonitique. *Soc. Franc. Bull. Mineral. Crist.,* 77:491–499.

Hey, M.H., 1954. A new review of the chlorites. *Mineral. Mag.,* 30:277–292.

Heystek, H., 1955. Some hydrous micas in South African clays and shales. *Proc. Natl. Conf. Clays Clay Miner. 3rd–Natl. Acad. Sci. Natl. Res. Counc., Publ.,* 395:337–355.

Heystek, H., 1962. Hydrothermal rhyolitic alteration in the Castle Mountains, California. *Clays Clay Miner., Proc.,* 11:158–168.

Heystek, H. and Schmidt, E.R., 1954. Palygorskite from Dornboom. *Trans. Geol. Soc. S. Afr.,* 56:99–115.

Hinckley, D.N., 1961. Mineralogical and chemical variations in the kaolin deposits of the Coastal Plain of Georgia and South Carolina. *Penn. State Univ. N.S.F. Tech. Rep.,* 180 pp.

Hofmann, U., Weiss, G.K., Mehler, A. and Scholz, A., 1956. Intracrystalline swelling, cation exchange, and anion exchange of minerals of the montmorillonite group and of kaolinite. *Proc. Natl. Conf. Clays Clay Miner. 4th – Natl. Acad. Sci. Natl. Res. Counc., Publ.,* 456:273–287.

Holmes, R.J., 1950. *Reference Clay Localities – Europe.* A.P.I. Proj. 49, Prelim. Rep., 4:101 pp.

Honess, A.P. and Williams, F.J., 1935. Dickite from Pennsylvania. *Am. Mineralogist,* 20:462–466.

Honjo, G., Kitamura, N. and Mihama, K., 1954. A study of clay minerals by means of single crystal electron diffraction diagrams: the structure of tubular kaolin. *Clay Miner.,* 2:133–141.

Hower, J., 1961. Some factors concerning the nature and origin of glauconite. *Am. Mineralogist,* 46:313–334.

Hower, J., 1967. Order of mixed-layering in illite/montmorillonites. *Clays Clay Miner., Proc.,* 15:63–74.

Hower, J. and Mowatt, T.C., 1966. The mineralogy of illites and mixed-layer illite/ montmorillonites. *Am. Mineralogist,* 51:825–854.

Hsu, P.H. and Bates, T.F., 1964. Fixation of hydroxy-aluminium polymers by vermiculite. *Proc. Soil. Sci. Soc. Am.*, 28:763–768.

Huggins, C.W., Denny, M.V. and Shell, H.R., 1962. Properties of palygorskite, an abestiform mineral. *U.S. Bur. Mines, Rept. Invest.*, 6071:8 pp.

Hurley, P.M., Cormier, J., Hower, H.W., Fairbairn, H.W. and Pinson Jr., W.H., 1960. Reliability of glauconites for age measurements by K-Ar and Rb-Sr methods. *Am. Assoc. Petrol. Geologists, Bull.*, 44:1793–1808.

Hutton, C.O. and Seelye, F.T., 1941. Composition and properties of some New Zealand glauconites. *Am. Mineralogist*, 26:595–604.

Iler, R.K., 1955. *The Colloid Chemistry of Silica and Silicates*. Cornell Univ. Press, New York, N.Y., 324 pp.

Jackson, M.L., 1963. Interlayering of expansible layer silicates in soils by chemical weathering. *Clays Clay Miner., Proc.*, 13:29–46.

Jaffe, H.W. and Sherwood, A.M., 1950. Phosphate-allophane in an epidosite from North Carolina. *Am. Mineralogist*, 35:102–107.

Kardymowicz, I., 1960. Celadonite from Barcza in the Swietokrzyskie Mountains (central Poland). *Kwart. Geol.*, 4:609–616.

Kauffman Jr., A.J., 1943. Fibrous sepiolite from Yavapai County, Arizona. *Am. Mineralogist*, 28:512–520.

Kautz, K., 1965. Zwei Seladonite ungewöhnlicher chemischer Zusammensetzung. *Beitr. Mineral. Petrogr.*, 11:398–404.

Keller, W.D., 1958. Glauconite mica in the Morrison Formation in Colorado. *Proc. Natl. Conf. Clays Clay Miner., 5th–Natl. Acad. Sci. Natl. Res. Counc., Publ.*, 566:120–128.

Keller, W.D., 1961. The origin of high-alumina clay minerals – a review. *Clays Clay Miner., Proc.*, 12:129–151.

Keller, W.D., 1963. Hydrothermal kaolinization (endellitization) of volcanic glassy rock. *Clays Clay Miner., Proc.*, 10:333–343.

Keller, W.D., 1964. Process of origin and alteration of clay minerals. In: *Soil Clay Mineralogy*. Univ. North Carolina Press.

Keller, W.D., Hanson, R.F., Huang, W.H. and Cervantes, A., 1971. Sequential active alteration of rhyolitic volcanic rock to endellite and a precursor phase of it at a spring in Michoacan, Mexico. *Clays Clay Miner.*, 19:121–127.

Kepezhinskas, K.B. and Sobolev, V.S., 1965. Paragenetic types of chlorite. *Dokl. Akad. Nauk S.S.S.R.*, 161:118–121. (Translated from: *Dokl. Akad. Nauk S.S.S.R.*, 161(2):436–439.)

Kerns Jr., R.L. and Mankin, C.J., 1967. Compositional variation of a vermiculite with particle size. *Clays Clay Miner., Proc.*, 15:163–177.

Kerr, P.F., 1950. Analytical data on reference clay materials. (Prelim. Rep., 7: *Reference Clay Minerals.*) Am. Petroleum Inst., Columbia Univ., New York, N.Y.

Kitagawa, Y., 1971. The "unit particle" of allophane. *Am. Mineralogist*, 56:465–475.

Kittrick, J.A., 1970. Precipitation of kaolinite at $25°C$ and 1 atmosphere. *Clays Clay Miner.*, 18:261–267.

Klages, M.G. and White, J.L., 1957. A chlorite-like mineral in Indiana soils. *Proc. Soil Sci. Soc. Am.*, 21:16–20.

Knechtel, M.M. and Patterson, S.H., 1962. Bentonite deposits of the northern Black Hills district, Wyoming, Montana, and South Dakota. *U.S. Geol. Surv., Bull.*, 1082-M:893–1030.

Kodama, H., 1966. The nature of the component layers of rectorite. *Am. Mineralogist*, 51:1035–1055.

Koenig, G.A., 1912. New observations in chemistry and mineralogy. *J. Phila. Acad. Nat. Sci.*, 15:407–426.

Konta, J., 1969. Raw kaolin from Osmosa-Bozícany deposits in west Bohemia. *Interceram*, 59–60:102–104; 115–117.

Konta, J. and Sindelar, J., 1955. Saponite from the fissure fillings of the amphibolites of Caslav. *Univ. Carolina (Prague), Geol.*, 1:177–186.

Köster, H.M., 1960. Nontronite und Picotit aus dem Basalt des Okberges bei Hundsangen, Westerwald. *Beitr. Mineral. Petrogr.*, 7:71–77.
Köster, H.M., 1965. Glaukonit aus der Regensburger Oberkreideformation. *Beitr. Mineral. Petrogr.*, 11:614–620.
Kuliev, A., 1959. Sauconite from the Kugitang Range. *Tr. Inst. Geol. Akad. Nauk Turkmen S.S.R.*, 2:161–165 (Chem. Abstr., 55:25614).
Kunze, G.W. and Bradley, W.F., 1964. Occurrence of a tubular halloysite in a Texas soil. *Clays Clay Miner., Proc.*, 12:523–527.
Kvalvaser, I.A., 1953. On celadonite from karadagh in the Crimea. *Mineral. Sb. Lvov. Geol. Soc.*, 7:223–226.
Lacroix, A., 1916. Sur le minéral colorant le plasma de Madagascar et sur la celadonite. *Soc. Franc. Minéral. Bull.*, 39:90–95.
Lacroix, A., 1941. Les gisements de phlogopite de Madagascar et les pyroxénites qui les renferment. *Ann. Géol. Serv. Mines Madagascar*, 11:1–119.
Langston, R.B. and Pask, J.A., 1969. The nature of anauxite. *Clays Clay Miner.*, 16:425–436.
Lazarenko, E.K., 1940. Donbassites, a new group of minerals from the Donetz Basin. *Compt. Rend. Acad. Sci. U.S.S.R.*, 28:519–521.
Lazarenko, E.K., 1956. Celadonites from the basalts of Volhyria. *Mineral. Sb. Lvov. Geol. Soc.*, 10:352–356.
Levi, M.G., 1914. Sulle celadoniti di alcune localita venet. *Rev. Mineral. Crist. Ital.*, 43:74.
Levinson, A.A., 1955. Relationship between polymorphism and composition in muscovite-lepidolite series. *Am. Mineralogist*, 40:41–49.
Linares, J. and Huertas, F., 1971. Kaolinite: synthesis at room temperature. *Science*, 171:896–897.
Lippman, F., 1954. Uber einen Keuperton von Zaisersweiker bei Maulbronn. *Heidelberger Beitr. Mineral. Petrogr.*, 4:130–134.
Lister, J.S. and Bailey, S.W., 1967. Chlorite polytypism, 4. Regular two-layer structures. *Am. Mineralogist*, 52:1614–1631.
Lodding, W., 1965. Kaolinite macrocrystals from New Woodstown, New Jersey. *Am. Mineralogist*, 50:1113–1114.
Longchambon, H. and Morgues, F., 1927. Sur le gisement de magnesite de Salinelles (Gard). *Bull. Soc. Franc. Minéral.*, 50:66–74.
Loughnan, F.C., 1957. A technique for the isolation of montmorillonite and halloysite. *Am. Mineralogist*, 42:393–398.
Loughnan, F.C., 1959. Further remarks on the occurrence of palygorskite at Red Banks Plains, Queensland. *Proc. R. Soc. Queensland*, 71:43–50.
MacEwan, D.M.C., 1947. The nomenclature of the halloysite minerals. *Mineral. Mag.*, 28:36–44.
MacEwan, D.M.C., 1954. "Cardenite" a trioctahedral montmorillonite derived from biotite. *Clay Miner.*, 2:120.
MacEwan, D.M.C., Ruiz, A.A. and Brown, G., 1961. Interstratified clay minerals. In: G. Brown (Editor), *The X-ray Identification and Crystal Structures of Clay Minerals*. London Mineral. Soc., London, pp. 393–445.
Mackenzie, R.C., 1957a. Saponite from Allt Ribhein, Fiskavaig Bay, Skye. *Mineral. Mag.*, 31:672–680.
Mackenzie, R.C., 1957b. The illite in some Old Red Sandstone soils and sediments. *Mineral. Mag.*, 31:681–689.
Mackenzie, R.C., 1960. The evaluation of clay mineral composition with particular reference to smectites. *Silicates Ind.*, 25:12–18; 71–75.
Mackenzie, R.C., 1963. Retention of exchangeable ions by montmorillonites. *Int. Clay Conf., Stockholm*: 183–193.
Mackenzie, R.C., Walker, G.F. and Hart, R., 1949. Illite occurring in decomposed granite at Ballater, Aberdeenshire. *Mineral. Mag.*, 28:704–714.

McAtee, J.L., 1958. Heterogeneity in montmorillonite. *Proc. Natl. Conf. Clays Clay Miner. 5th–Natl. Acad. Sci. Natl. Res. Counc., Publ.*, 566:279–288.

McConnell, D., 1954. An American occurrence of volkenskoite. *Proc. Natl. Conf. Clays Clay Miner., 2nd–Natl. Acad. Sci. Natl. Res. Counc., Publ.*, 327:152–157.

McLaughlin, R.J.W., 1959. The geochemistry of some kaolinitic clays. *Geochim. Cosmochim. Acta*, 17:11–16.

Mägdefrau, E. and Hofmann, U., 1937. Glimmertige Mineralien als Tonsubstanzen. *Z. Kryst.*, 98:31–59.

Maksimovic, Z. and Radukii, G., 1961. Sepiolite from Goles near Lipljan (south Serbia). *Am. Geol. Peninsula Balkan*, 28:309–316.

Malden, P.J. and Meads, R.E., 1967. The solid state. *Nature*, 215:844–845.

Maleev, E.F. and Maleev, E.O., 1949. On the Baranovsky Volcano in the Amur-Ussuri Depression. *Trans. Volcan. Lab. Kamchatka Volcan. Sta.*, 6: 23–52.

Malkova, K.A., 1956. Celadonite from the Bug region. *Mineral. Sb. Lvov. Geol. Soc.*, 10:305–318.

Manghnani, M.H. and Hower, J., 1964. Glauconites: cation exchange capacities and infrared spectra. *Am. Mineralogist*, 49:586–598.

Mankin, C.J. and Dodd, C.G., 1963. Proposed reference illite from the Ouachita Mountains of southeastern Oklahoma. *Clays Clay Minerals, Proc.*, 10:373–379.

Martin-Vivaldi, J.L. and Cano-Ruiz, J. 1955. Contribution to the study of sepiolite, 2. Some considerations regarding the mineralogical formula. *Proc. Natl. Conf. Clays Clay Miner., 4th – Natl. Acad. Sci. Natl. Res. Counc., Publ.*, 456:166–172.

Mathieson, A.M., 1958. Mg-vermiculite: a refinement and re-examination of the crystal structure of the 14.36Å phase. *Am. Mineralogist*, 43:216–227.

Mathieson, A.M. and Walker, G.F., 1954. Structure of Mg-vermiculite. *Am. Mineralogist*, 39:231–255.

Maxwell, G.T. and Hower, J., 1967. High-grade diagenesis and low-grade metamorphism of illite in the Precambrian Belt. *Am. Mineralogist*, 52:843–857.

Mehra, O.P. and Jackson, M.L., 1959. Constancy of the sum of mica unit cell potassium surface and interlayer sorption in vermiculite-illite clays. *Proc. Soil Sci. Soc. Am.*, 23:101–105.

Meyer, D.B., 1935. A sericite of unusual composition. *Am. Mineralogist*, 20:384–388.

Michalek, Z. and Stock, L., 1958. Allophane from the Carpathian Flysch. *Bull. Acad. Polon. Sci. Ser., Chim. Geol. Geogr.*, 6:384–388.

Midgley, H.G., 1959. A sepiolite from Mullion, Cornwall. *Clay Mineral., Bull.*, 4: 88–93.

Millot, G., 1949. Relations entre la constitution et la genèse des roches sédimentaires argileuses. *Géol. Appl. Prosp. Min.*, 2.

Millot, G., 1953. Héritage et néoformation dans la sédimentation argileuse. *Compt. Rend., Congr. Géol. Int.*, 18:163–217.

Millot, G., 1964. *Géologie des Argiles*. Masson, Paris.

Millot, G. and Camez, T., 1963. Genesis of vermiculite and mixed-layered vermiculite in the evolution of the soils of France. *Clays Clay Miner., Proc.*, 10:90–95.

Minato, H. and Takano, Y., 1952. On the iron sericites and magnesium sericite from Unnan mine, Shimani prefecture, Japan. *Sci. Pap. Coll. Gen. Educ., Univ. Tokyo*, 2:189.

Miyamoto, N., 1957. Iron-rich saponite from Mazé, Nigata prefecture, Japan. *Mineral. J. Japan*, 2:193–195.

Mongiorgi, R. and Morandi, N., 1970. Al saponite e strati misti clorite-Al saponite nelle idrotermaliti di una breccia a contatto coi diabasi di Rossena nell'Appennino reggiano. *Mineral. Petrogr. Acta*, 16:139–154.

Muir, A., 1951. Notes on the soils of Syria. *J. Soil Sci.*, 12:168–181.

Müller, A., 1961. "Al-chlorite" a new dioctahedral mineral of the chlorite group. *Clays Clay Miner. Abstr.*

Müller, A., 1963. Zur Kenntnis di-oktaedrischer Vierschichtphyllosilikate (Sudoit-Reihe der Sudoit-Chlorit-Gruppe). *Int. Clay Conf., 1st, Proc.*, 1:121–130.

Mumpton, F.A. and Roy, R., 1958. New data on sepiolite and attapulgite. *Proc. Natl. Conf. Clays Clay Miner. 5th–Natl. Acad. Sci. Natl. Res. Counc., Publ.*, 566:136–143.
Nagelschmidt, G. and Hicks, D., 1943. The mica of certain coal-measure shales in south Wales. *Mineral. Mag.*, 26:297–303.
Nagelschmidt, G., Donnelly, H.F. and Morcom, A.J., 1949. On the occurrence of anatase in sedimentary kaolin. *Mineral. Mag.*, 28:492–495.
Nagy, B. and Bradley, W.F., 1955. Structure of sepiolite. *Am. Mineralogist*, 40:885–892.
Nelson, B.W. and Roy, R., 1958. Synthesis of the chlorites and their structural and chemical constitution. *Am. Mineralogist*, 43:707–725.
Newnham, R.E. and Brindley, G.W., 1956. The crystal structure of dickite. *Acta Cryst.*, 9:759–764.
Novák, F. and Valcha, Z., 1964. Ferric orthochamosite from Hora Svate Katřiny in the Krušné Hory Mountains. *Sb. Geol. Věd Tech. Geochem.*, 3:7–27.
Okamoto, G., Okura, T. and Goto, K., 1957. Properties of silica in water. *Geochim. Cosmochim. Acta*, 12:123–132.
Ormsby, W.C. and Sand, L.B., 1954. An analytical tool for mixed-layer aggregates. *Proc. Natl. Conf. Clays Clay Miner., 2nd–Natl. Acad. Sci. Natl. Res. Counc., Publ.*, 327:254–263.
Osthaus, B.B., 1954. Chemical determination of tetrahedral ions in nontronite and montmorillonite. *Proc. Natl. Conf. Clays Clay Miner. 2nd–Natl. Acad. Sci. Natl. Res. Counc., Publ.*, 327:404–417.
Osthaus, B.B., 1955. Interpretation of chemical analyses of montmorillonites. *Conf. Clays Clay Tech., 1st*: 95–100.
Ovcharenko, F.D., 1964. *The Colloid Chemistry of Palygorskite*. Acad. Sci. of the Ukranian S.S.R., 101 pp. (Translated by Israel Program for Scientific Translations, Jerusalem.)
Owens, J.P. and Minard, J.P., 1960. Some characteristics of glauconite from the coastal plain formations of New Jersey. *U.S. Geol. Surv., Prof. Pap.*, 400-B:430–432.
Oyawaye, M.O. and Hirst, D.M., 1964. Occurrence of a montmorillonite mineral in the Ingerian Younger Granites at Ropp, Plateau Province, northern Nigeria. *Clay Miner.*, 5:427–433.
Parry, W.T., 1968. Sepiolite from Pluvial Mound Lake, Lynn and Terry Counties, Texas. *Am. Mineralogist*, 53:984–993.
Parry, W.T. and Reeves, C.C., 1966. Lacustrine glauconitic mica from Pluvial Lake Mound, Lynn and Terry Counties, Texas. *Am. Mineralogist*, 51:229–235.
Parry, W.T. and Reeves, C.C., 1968. Sepiolite from Pluvial Mound Lake, Lynn and Terry Counties, Texas. *Am. Mineralogist*, 53:984–993.
Patterson, S.H., 1963. Halloysitic underclay and amorphous inorganic matter in Hawaii. *Clays Clay Miner., Proc.*, 12:153–172.
Patterson, S.H. and Hosterman, J.W., 1962. Geology and refractory clay deposits of the Haldeman and Wrigley Quadrangles, Kentucky. *U.S. Geol. Surv., Bull.*, 1122-F:113 pp.
Pauling, L., 1960. *The Nature of the Chemical Bond*. Cornell Univ. Press, New York, N.Y., 3rd ed., 449 pp.
Perry, E. and Hower, J., 1970. Burial diagenesis in Gulf Coast pelitic sediments. *Clays Clay Miner.*, 18:165–177.
Peterson, N.M.A., 1961. Expandable chloritic clay minerals from Upper Mississippian carbonate rocks, of the Cumberland Plateau in Tennessee. *Am. Mineralogist*, 46:1245–1269.
Pirani, R., 1963. Sul fillosilicato dei livelli eruttivi di Monte Bonifato di Alcamo e di Monte Barbaro di Segesta e sulla validità di uso della nomenclatura binomia: glauconite-celadonite. *Mineral. Petrogr. Acta*, 9:31–78.
Pollard Jr., C.O., 1971. Semidisplacive mechanism for diagenetic alteration of montmorillonite layers to illite layers. Appendix in C.E. Weaver and K.C. Beck, Clay-water diagenesis during burial: how mud becomes gneiss. *Geol. Soc. Am., Spec. Pap.*, 134:79–93.
Poncelet, G. and Brindley, G.W., 1967. Experimental formation of kaolinite from montmorillonite at low temperature. *Am. Mineralogist*, 53:1161–1173.

Porrenga, D.H., 1966. Clay minerals in Recent sediments of the Niger Delta. *Clays Clay Minerals, Proc.*, 14:221–233.

Porrenga, D.H., 1968. Non-marine glauconitic illite in the Lower Oligocene of Aardenburg, Belgium. *Clay Minerals*, 7:421–430.

Powers, M.C., 1957. Adjustment of land-derived clays to the marine environment. *J. Sediment. Petrol.*, 20:355–372.

Powers, M.C., 1959. Adjustment of clays to chemical change and the concept of the equivalence level. *Clays Clay Miner., Proc. 6th Conf.*, pp. 309–326.

Preisinger, A., 1959. An X-ray study of the structure of sepiolite. *Clays Clay Miner., Proc. 6th Conf.*, pp. 61–67.

Quigley, R.M. and Martin, R.T., 1963. Chloritized weathering products of a New England glacial till. *Clays Clay Minerals, Proc.*, 10:107–116.

Radonova, T.G. and Karadzhova, B., 1966. Celadonites from the Sredna Bora area and their genesis. *Chem. Abstr.*, 65: 10345 (Translated from *Tr. Vurkhu Geol. Bulgar., Ser. Geokhim., Mineral. Petrogr., Bulgar. Akad Nauk*, 6:193–206.)

Radoslovich, E.W., 1962. The cell dimensions and symmetry of layer-lattice silicates, 2. Regression relations. *Am. Mineralogist*, 47:617–636.

Radoslovich, E.W., 1963a. The cell dimensions and symmetry of layer-lattice silicates, 4. Interatomic forces. *Am. Mineralogist*, 48:76–99.

Radoslovich, E.W., 1963b. The cell dimensions and symmetry of layer-lattice silicates, 5. Composition limits. *Am. Mineralogist*, 48:348–367.

Radoslovich, E.W., 1963c. The cell dimensions and symmetry of layer-lattice silicates, 6. Serpentine and kaolin morphology. *Am. Mineralogist*, 48:368–378.

Radoslovich, E.W. and Norrish, K., 1962. The cell dimensions and symmetry of layer-lattice silicates, 1. Some structural considerations. *Am. Mineralogist*, 47:599–616.

Raman, K.V. and Jackson, M.L., 1966. Layer charge relations in clay minerals of micaceous soils and sediments. *Clays Clay Minerals, Proc.*, 14:53–68.

Ramberg, H., 1952. Chemical bonds and distribution of cations in silicates. *J. Geol.*, 60:331–355.

Rateev, M.A., 1963. Modification degree of clay minerals during the stage of sedimentation and diagenesis of marine deposits. *Int. Clay Conf., Stockholm*, 2:171–180.

Rateyev, M.A., Gradusov, B.P. and Kheirov, M.B., 1969. Potassium rectorite from the Upper Carboniferous of the Samarskaya Luka (Samara Bend of the Volga). *Dokl. Akad. Nauk S.S.S.R.*, 185:116–119.

Raup, O.B., 1966. Clay mineralogy of Pennsylvanian redbeds and associated rocks flanking ancestral Front Range of central Colorado. *Am. Assoc. Petrol. Geologists, Bull.*, 50:251–268.

Reesman, A.L. and Keller, W.D., 1967. Chemical composition of illite. *J. Sediment. Petrol.*, 37:592–596.

Rex, R.W., 1967. Authigenic silicates formed from basaltic glass by more than 60 million years contact with sea water, Sylvania Guyot, Marshall Islands. *Clays Clay Miner., Proc.*, 15:195–203.

Reynolds, R.C., 1963. Potassium-rubidium ratios and polymorphism in illites and microclines from the clay-size fractions of Proterozoic carbonate rocks. *Geochim. Cosmochim. Acta*, 27:1097–1112.

Reynolds, R.C., 1965. Geochemical behavior of boron during the metamorphism of carbonate rocks. *Geochim. Cosmochim. Acta*, 29:1101–1114.

Rich, C.I., 1958. Muscovite weathering in a soil developed in the Virginia piedmont. *Proc. Natl. Conf. Clays Clay Miner., 5th–Natl. Acad. Sci. Natl. Res. Counc., Publ.*, 566:203–212.

Rich, C.I., 1968. Hydroxy interlayers in expansible layer silicates. *Clays Clay Miner.*, 16:15–30.

Rich, C.I. and Cook, M.G., 1963. Formation of dioctahedral vermiculite in Virginia soils. *Clays Clay Miner., Proc.*, 10:96–106.

Rich, C.I. and Obenshain, S.S., 1955. Chemical and clay-mineral properties of red-yellow podzolic soil derived from muscovite schist. *Proc. Soil Sci. Soc. Am.*, 19:334–339.

Roberson, H.E., 1964. Petrology of Tertiary bentonites of Texas. *J. Sediment. Petrol.*, 34:401–411.

Roberson, H.E., Weir, A.H. and Woods, R.D., 1968. Morphology of particles in size-fractionated Na-montmorillonites. *Clays Clay Miner.*, 16:239–247.

Robertson, R.H.S., 1963. Allophanic soil from Trail Bridge, Oregon, with notes on mosaic growth in clay minerals. *Clay Miner.*, 5:237–247.

Robertson, R.H.S., Brindley, G.W. and Mackenzie, R.C., 1954. Mineralogy of kaolin clays from Puger, Tanganiyika. *Am. Mineralogist*, 39:118–138.

Rogers, L.E., Martin, A.E. and Norrish, K., 1954. Palygorskite from Queensland. *Mineral. Mag.*, 30:534–540.

Rogers, L.E., Quirk, J.P. and Norrish, K., 1956. Aluminum sepiolite. *J. Soil Sci.*, 7:177–183.

Rohrlich, V., Price, N.B. and Calvert, S.E., 1969. Chamosite in recent sediments of Loch Etive, Scotland. *J. Sediment. Petrol.*, 39:624–631.

Ronov, A.B. and Khlebnikova, Z.V., 1957. Chemical composition of the main genetic clay types. *Geokhimiya*, 6:449–469. (Translation in *Geochemistry*, 6:527–552.)

Ross, C.S., 1946. Sauconite – a clay mineral of the montmorillonite group. *Am. Mineralogist*, 31:411–424.

Ross, C.S., 1960. Review of the relations in the montmorillonite group of clay minerals. *Clays Clay Miner. Proc. 7th Conf.*, pp. 225–229.

Ross, C.S. and Hendricks, S.B., 1945. Minerals of the montmorillonite group. *U.S. Geol. Surv., Prof. Pap.*, 205-B: 79 pp.

Ross, C.S. and Kerr, P.F., 1931. The kaolin minerals. *U.S. Geol. Surv., Prof. Pap.*, 165-E:151–176.

Ross, C.S. and Kerr, P.F., 1934. Halloysite and allophane. *U.S. Geol. Surv., Prof. Pap.*, 185-G: 135–148.

Roth, C.B., Jackson, M.L. and Syers, J.K., 1969. Deferration effect on structural ferrous-ferric iron ratio and C.E.C. of vermiculites and soils. *Clays Clay Miner.*, 17:253–264.

Roy, D.M. and Roy, R., 1954. An experimental study of the formation and properties of synthetic serpentines and related layer silicate minerals. *Am. Mineralogist*, 39:957–975.

Roy, R. and Romo, L.A., 1957. Weathering studies, 1. New data on vermiculite. *J. Geol.*, 65:603–610.

Ruotsala, A.P., Pfluger, C.E. and Garnett, M., 1964. Iron-rich serpentine and chamosite from Ely, Minnesota. *Am. Mineralogist*, 49:993–1001.

Sand, L.B., 1956. On the genesis of residual kaolins. *Am. Mineralogist*, 41: 28–40.

Sawhney, B.L., 1958. Aluminum interlayers in soil clay minerals, montmorillonite and vermiculite. *Nature*, 182:1595–1596.

Sawhney, B.L. and Jackson, M.L., 1958. Soil montmorillonite formulas. *Proc. Soil Sci. Soc. Am.*, 22:115–118.

Schaller, W.T., 1950. An interpretation of the composition of high-silica sericites. *Mineral. Mag.*, 29:406.

Schellmann, W., 1966. Secondary formation of chamosite from goethite. *Z. Erzbergbau Metall-hüttenw.*, 19(6):302–305.

Scherillo, A., 1938. Celadonite in an eruptive rock in Erythrea. *Period Miner. Roma*, 9:253–264.

Schmidt, E.R. and Heckroodt, R.O., 1959. A dickite with an elongated crystal habit and its dehydroxylation. *Mineral. Mag.*, 33:314–323.

Schneider, H., 1927. A study of glauconite. *J. Geol.*, 35:289–310.

Schoen, R., 1964. Clay minerals of the Silurian Clinton ironstones, New York State. *J. Sediment. Petrol.*, 34:855–863.

Schofield, R.K., 1949. Effect of pH on electric charges carried by clay particles. *J. Soil Sci.*, 1:1–8.

Schofield, R.K. and Samson, H.R., 1953. The deflocculation of kaolinite suspensions and the accompanying change over from positive to negative chloride adsorption. *Clay Miner.*, 2:45–50.

Schultz, L.G., Shepard, A.O., Blackmon, P.D. and Starkey, H.C., 1969. Mixed-layer kaolinite-montmorillonite from the Yucatan Peninsula. *Clay Miner. Soc. Abstr., 18th Conf.*, p. 33.

Sedletsky, I.D., 1937. Genesis of minerals from soil colloids of the montmorillonite group. *Compt. Rend. Acad. Sci. U.S.S.R.*, 17:375–377.

Seed, D.P., 1965. The formation of vermicular pellets in New Zealand glauconites. *Am. Mineralogist*, 50:1097–1106.
Serdyuchenko, D.P., 1933. Chrome-nontronites and their genetical relations with the serpentines of the northern Caucasus. *Mem. Soc. Russe Mineral., Ser. 2*, 62:376–390.
Serdyuchenko, D.P., 1953. Magnesium parahalloysites and other montmorillonite minerals from Jurassic sediments of the northern Caucasus. *Vopr. Petrogr. Mineral., Akad. Nauk S.S.S.R.*, 2:100–122.
Shannon, E.V., 1926. The minerals of Idaho. *U.S. Nat. Mus. Bull.*, 131:367.
Shaskina, V.P., 1961. Mineralogy of weathered basalt crust in west Volynya. *Int. Geol. Rev.*, 393–407. (Translated from *Lvov. Geol. Obshch. Mineralog. Sb.*, 13:190–211; in Russian.)
Shaw, D.M., 1961. Manipulation errors in geochemistry: a preliminary study. *Trans. R. Soc. Can., 3rd Ser.*, 55:41–55.
Shimoda, S., 1969. New data for tosudite. *Clays Clay Minerals*, 17:179–184.
Shirozu, H., 1958. X-ray powder patterns and cell dimensions of some chlorites in Japan, with a note on their interference colors. *Mineral. J. Japan*, 2: 209–223.
Shirozu, H. and Bailey, S.W. 1965. Chlorite polytypism, 3. Crystal structure of an orthohexagonal iron chlorite. *Am. Mineralogist*, 50:868–885.
Shiveley Jr., R.R. and Weyl, W.A., 1951. The color change of ferrous hydroxide on oxidation. *J. Phys. Colloid Chem.*, 55:512–515.
Siffert, B., 1962. Quelques réactions de la silice en solution. La formation des argiles. *Mém. Serv. Carte Géol. Alsac. Lorraine*, 21.
Slaughter, M. and Milne, I.H., 1958. The formation of chlorite-like structures from montmorillonite. *Clays Clay Miner., Proc. 7th Conf.*, pp. 114–124.
Smith, J.P. and Brown, W.E., 1959. X-ray studies of aluminum and iron phosphates containing potassium or ammonium. *Am. Mineralogist*, 44:138–142.
Smith, R.W., 1929. Sedimentary kaolins of the Coastal Plain of Georgia. *Geol. Surv. Georgia, Bull.*, 44:482 pp.
Smithson, F. and Brown, G., 1954. The petrography of dickite sandstones in north Wales and northern England. *Geol. Mag.*, 91:177–188.
Smithson, F. and Brown, G., 1957. Dickite from sandstone in northern England and north Wales. *Mineral. Mag.*, 31:381–391.
Smulikowski, K., 1936. On skolite, a new mineral of the glauconite group. *Arch. Mineral.*, 12:144–178.
Smulikowski, K., 1954. The problem of glauconite. *Polska Akad. Nauk, Kom. Geol. Arch. Mineral. Warsz.*, 18(1):21–120.
Spangenberg, K., 1938. Die hydroxydischen Nickel- und Magnesiasilikat-Mineralien. *Naturwissenschaften*, 26:578–579.
Spiro, N.S. and Granberg, I.S., 1964. Composition of the complex adsorbed by argillaceous rocks as an indicator of conditions during the early stage of sedimentation. *Int. Geol. Rev.*, 6:1076–1079.
Steinfink, H., 1958a. The crystal structure of chlorite, 1. A monoclinic polymorph. *Acta Cryst.*, 11:191–195.
Steinfink, H., 1958b. The crystal structure of chlorite, 2. A triclinic polymorph. *Acta Cryst.*, 11:195–198.
Stephen, I., 1954. Polygorskite from Shetland. *Mineral. Mag.*, 30:471–480.
Stout, P.P., 1940. Alterations in the crystal structure of the clay minerals as a result of phosphate fixation. *Proc. Soil Sci. Soc. Am.*, 4:177–182.
Sudo, T., 1954. Iron-rich saponite found from Tertiary iron sand beds of Japan: re-examination of "lembergite", *J. Geol. Soc. Japan*, 60:18–27.
Sudo, T. and Hayashi, H., 1956. A randomly interstratified kaolin-montmorillonite in acid clay deposits in Japan. *Nature*, 178:1115–1116.
Sudo, T. and Kodama, H., 1957. An aluminian mixed-layer mineral of montmorillonite-chlorite. *Z. Krist.*, 109:380–387.

Sudo, T. and Sato, M. 1966. Dioctahedral chlorite. *Proc. Int. Clay Conf., Israel,* pp. 33–40.
Sudo, T. and Takahashi, H., 1956. Shapes of halloysite particles in Japanese clays. *Proc. Natl. Conf. Clays Clay Miner., 4th–Natl. Acad. Sci. Natl. Res. Counc., Publ.,* 456:67–79.
Sumner, M.E., 1963. Effect of iron oxides on positive and negative charges in clays and soils. *Clay Miner.,* 5:218–224.
Swindale, L.D. and Fan, Pow-Foong, 1967. Transformation of gibbsite to chlorite in ocean bottom sediments. *Science,* 157:799–800.
Swineford, A., McNewt, J.D. and Crumpton, C.F., 1954. Hydrated halloysite in Blue Hill Shale. *Proc. Natl. Conf. Clays Clay Miner. 2nd–Natl. Acad. Sci. Natl. Res. Counc., Publ.,* 327:158–170.
Takahashi, H., 1939. Synopsis of glauconization. In: *Recent Marine Sediments, a Symposium,* pp. 503–513.
Takahashi, H., 1958. Structural variations of kaolin minerals. *Bull. Chem. Soc. Japan,* 31:275–283.
Tamura, T., 1958. Identification of clay minerals from acid soils. *J. Soil Sci.,* 9:141–147.
Tamura, T. and Jackson, M.L., 1953. Structural and energy relationships in the formation of iron and aluminum oxides, hydroxides and silicates. *Science,* 117:331–383.
Taylor, G.L. Ruotsala, A.P. and Keeling Jr., R.O., 1968. Analysis of iron in layer silicates by Mössbauer spectroscopy. *Clays Clay Miner., Proc.,* 16:381–391.
Thomas, G.W. and Swoboda, A.R., 1963. Cation exchange in kaolinite-iron oxide system. *Clays Clay Miner., Proc.,* 11:321–326.
Thompson, G.R. and Hower, J., 1971. Some revisions of the crystal chemistry of glauconite. *Clay Miner. Soc. Abstr., 20th Conf.,* p.35.
Tien, Pei-Lin, 1969. A purple-colored 1M mica clay from Silverton, Colorado. *Clays Clay Miner.,* 17:245–249.
Triplehorn, D.M., 1967. Occurrence of pure, well-crystallized 1M illite in Cambro-Ordovician sandstone from Rhourde El Baguel Field, Algeria. *J. Sediment. Petrol.,* 37:879–884.
Turnock, A.C. and Eugster, H.P., 1958. Iron-rich chlorites. *Carnegie Inst. Wash. Yearbook, Ann. Rep., 1957–58.:* 191–192.
Tyler, S.A. and Bailey, S.W., 1961. Secondary glauconite in the Bivabic iron-formation of Minnesota. *Econ. Geol.,* 56:1030–1044.
Vakhrushev, V.A. and Baxpymeb, B.A. 1949. On the ferrihalloysite from the Anatolsky silicate-nickel ore deposit in the middle Urals. *Mém. Soc. Russe Mineral., Ser.,* 2, 78:272–274.
Van den Heuvel, R.C., 1966. The occurrence of sepiolite and attapulgite in the calcareous zone of a soil near Las Cruces, New Mexico. *Clays Clay Mineral. Proc.,* 13:193–207.
Van der Kaaden, G. and Quakernaat, J., 1968. Manganiferous smectite from Tirebolu (northeastern Turkey). *Contrib. Mineral. Petrol.,* 19:302–308.
Van der Marel, H.W., 1954. Potassium fixation in Dutch soils: mineralogical analyses. *Soil Sci.,* 78:163–179.
Van der Marel, H.W., 1958. Quantitative analysis of kaolinite. *J. Int. Etud. Argiles,* 1:1–19.
Velde, B., 1964. Mixed-layer mineral associations in muscovite-celadonite and muscovite. *Clays Clay Miner., Proc.,* 13:29–32.
Velde, B., 1965. Experimental determination of muscovite polymorph stabilities. *Am. Mineralogist,* 50:436–449.
Velde, B. and Hower, J., 1963. Petrologic significance of illite polymorphism in Paleozoic sedimentary rocks. *Am. Mineralogist,* 48:1239.
Veniale, F. and Van der Marel, H.W., 1968. A regular talc-saponite mixed-layer mineral from Ferriere, Nure Valley (Piacenza province, Italy). *Contrib. Mineral. Petrol.,* 17:237–254.
Von Knorring, O., Brindley, G.W. and Hunter, K., 1952. Nacrite from Hirvivaara, northern Karelia, Finland. *Mineral. Mag.,* 29:963–973.
Walker, G.F., 1949. The decomposition of biotite in the soil. *Mineral. Mag.,* 28:693–703.
Walker, G.F., 1950. Trioctahedral minerals in the soil clays of northeast Scotland. *Mineral. Mag.,* 29:72–84.

Walker, G.F., 1956. The mechanism of dehydration of Mg-vermiculite. *Proc. Natl. Conf. Clays Clay Miner. 4th–Natl. Acad. Sci. Natl. Res. Counc., Publ.,* 456:101–115.

Walker, G.F., 1957. On the differentiation of vermiculites and smectites in clays. *Clay Miner.,* 3:154–164.

Walker, G.F., 1958. Reaction of expanding-lattice minerals with glycerol and ethylene glycol. *Clay Miner.,* 3:302–313.

Walker, G.F., 1961. Vermiculite minerals. In: G. Brown (Editor), *The X-ray Identification and Crystal Structures of Clay Minerals.* London Mineral. Soc., London, pp. 297–324.

Warde, J.M., 1950. Refractory clays in the Union of South Africa. *J. Am. Ceram. Soc.,* 29:257–260.

Warshaw, C.M., 1957. *The Mineralogy of Glauconite.* Thesis, Pennsylvania State Univ.

Warshaw, C.M. and Roy, R., 1961. Classification and a scheme for the identification of layer silicates. *Geol. Soc. Am., Bull.,* 72:1455–1492.

Weaver, C.E., 1953. A lath-shaped non-expanded dioctahedral 2:1 clay mineral. *Am. Mineralogist,* 38:279–289.

Weaver, C.E., 1956. The distribution and identification of mixed-layer clays in sedimentary rocks. *Am. Mineralogist,* 41:202–221.

Weaver, C.E., 1958. The effects and geologic significance of potassium "fixation" by expandable clay minerals derived from muscovite, biotite, chlorite, and volcanic material. *Am. Mineralogist,* 43:839–861.

Weaver, C.E., 1959. The clay petrology of sediments. *Proc. Natl. Conf. Clays Clay Miner., 6th – Natl. Acad. Sci. Natl. Res. Counc., Publ.,* 566:154–187.

Weaver, C.E., 1960. Possible uses of clay minerals in search for oil. *Bull. Am. Assoc. Petrol. Geologists,* 44:1505–1518.

Weaver, C.E., 1961a. The Ouachita System, by P.T. Flawn, A. Goldstein Jr., P.B. King and C.E. Weaver. *Univ. Texas Publ.,* 6120.

Weaver, C.E., 1961b. Clay mineralogy of the Late Cretaceous rocks of the Washakie Basin. *Wyo. Geol. Assoc., Guidebook,* 16:145–152.

Weaver, C.E., 1964. Origin and significance of clays in sediments. In: *Petroleum Geochemistry,* Elsevier, Amsterdam, pp. 37–75.

Weaver, C.E., 1965. Potassium content of illite. *Science,* 147:603–605.

Weaver, C.E., 1967. Potassium, illite, and the ocean. *Geochim. Cosmochim. Acta,* 31:281–296.

Weaver, C.E., 1968. Electron microprobe study of kaolin. *Clays Clay Miner.,* 16:187–189.

Weaver, C.E. and Beck, K.C., 1971a. Clay water diagenesis during burial: how mud becomes gneiss. *Geol. Soc. Am., Spec. Pap.,* 134: 96 pp.

Weaver, C.E. and Beck, K.C., 1971b. Vertical variability in the attapulgite mining area. In: *Forum on the Geology of Industrial Mining.* Fla. Dept. Nat. Resources, in press.

Weaver, C.E., Wampler, J.M. and Pecuil, T.C., 1967. Mössbauer analysis of iron in clay minerals. *Science,* 156:504–508.

Weir, A.H. and Greene-Kelly, R., 1962. Beidellite. *Am. Mineralogist,* 47:137–146.

Weiss, A., Koch, G. and Hoffmann, U., 1955. Saponite. *Ber. Dtsch. Keram. Ges.,* 32:12–17.

Weiss, A., Scholz, A. and Hoffmann, U., 1956. Zur Kenntniss von Trioktaedrischen Illit. *Z. Naturforsch.,* 11:429–430.

Wells, R.C., 1937. Analyses of rocks and minerals from the laboratory of the U.S. Geological Survey. *U.S. Geol. Surv., Bull.,* 878.

Wermund, E.G., 1961. Glauconite in Early Tertiary sediments of Gulf Coastal Province. *Am. Assoc. Petrol. Geologists, Bull.,* 45: 1667–1697.

White, J.L., 1962. X-ray differation studies on weathering of muscovite. *Soil Sci.,* 93:16–21.

White, W.A., 1953. Allophanes from Lawrence County, Indiana. *Am. Mineralogist,* 38:634–642.

Whitehouse, U.G. and McCarter, R.S., 1958. Diagenetic modification of clay mineral types in artificial sea water. *Proc. Natl. Conf. Clays Clay Miner., 5th–Natl. Acad. Sci. Natl. Res. Counc., Publ.,* 566:81–119.

REFERENCES

Wiersma, J., 1970. Provenance, genesis and paleo-geographical implications of microminerals occurring in sedimentary rocks of the Jordan Valley area. *Publ. Fys. Geogr. Bodemk. Lab. Univ. Amsterdam.*

Willets, W.R. and Marchetti, F.R., 1958. Titanium pigments. *TAPPI Mono. Ser.*, 20:196–223.

Wilson, M.J., Bain, D.C. and Mitchell, W.A., 1968. Saponite from the Dalradian meta-limestones of northeast Scotland. *Clay Minerals*, 7:343–349.

Wise, W.S. and Eugster, H.P., 1964. Celadonite: synthesis, thermal stability and occurrences. *Am. Mineralogist*, 49:1031–1083.

Worrall, W.E. and Cooper, A.E., 1966. Ionic composition of a disordered kaolinite. *Clay Miner.*, 6:341–344.

Yarzhemskii, Ya. Ya., 1949. Petrographic character of recent salt sediments (in Russia). *Dokl. Akad. Nauk S.S.S.R.*, 68: 1085–1088. (*Chem. Abstr.*, 44:2901b).

Yoder, H.S., 1959. Experimental studies of micas: a review. *Clays Clay Miner., Proc. 6th Conf.*, pp. 42–60.

Yoder, H.S. and Eugster, H.P., 1955. Synthetic and natural muscovites. *Geochim. Cosmochim. Acta*, 8:225–280.

Yoshinaga, N. and Aomine, S., 1962. Imogolite in some Ando soils. *Soil Sci. Plant Nutr. (Tokyo)*, 8:22–29.

Youell, R.F., 1955. Mineralogy and crystal structure of chamosite. *Nature*, 176:560–561.

Youell, R.F., 1958. Isomorphous replacement in the kaolin group of minerals. *Nature*, 181:557–558.

Zussman, J., 1954. Investigation of the crystal structure of antigorite. *Mineral Mag.*, 30:498–512.

Zvyagin, B.B., 1957. Determination of the structure of celadonite by electron diffraction. *Kristallaografiya*, 2:388–394.

INDEX

Allevardite, 107
Allophane, 155–158
– cation exchange capacity, 157
–, chemical analyses, 156
–, morphological series with kaolinite, 154
–, morphology, 157, 158
– occurrence, 155
– ordering, 155
– origin, 155
–, phosphate, 157
– water, adsorbed, 157
– –, structural, 157
Alunite, 151
Amesite, 89
–, b-axis displacements, 167
–, chemical analysis, 164
– octahedral sheet, composition of, 165, 173
–, structural formula, 164
– synthesis, 173
Amorphous material, 69
– – in kaolinite,
– –, synthesis of, 169
Antigorite, 1
Apatite, 38
–, constituent in glauconite, 31, 37
Attapulgite, 4, 119, 124–126
– cation exchange capacity, 123, 125, 178
–, chemical analyses, 123
– composition, compared to illite, 186
– –, compared to montmorillonite, 186
– exchangeable cations, 125
– occurrence, 125
– octahedral sites, composition, 125, 174, 176, 177
– origin, 125, 126
– structure, 119
– –, related to composition, 183, 186, 187
– water, bound, 124, 125
– –, structural, 124, 125
– –, zeolitic, 124, 125
Anatase, 136
Anauxite, 131
–, chemical analyses, 132

Antigorite, 159

Ball clay, 142
Bayerite, 171
Beidellite, 3
–, chemical analyses, 60
–, definition of, 63
– layer charge related to composition, 179
– octahedral sheet, composition, compared to glauconite, 29
–, structural formulas, 60
–, tetrahedral rotation, 184
Bentonite, constituents of, 73, 113
– exchangeable cations, 71
Berthierine, 160
Biotite, 2, 173
– cation distribution, 173
– interlayer cations, 30
–, leached, 3
Biotite-illite, 18
Biotite-vermiculite, 106
Boehmite, 171
Brammallite, 18
Brucite sheet, 1
–, mixed-layer chlorite-montmorillonite, 116, 117
– stacking, 89, 90
 synthesis, 170
–, weathering of (chlorite), 92
Brunsvigite, 87

Cardenite, 83
Celadonite, 3, 47–53
– cation distributions, 50
–, chemical analyses, 48, 49
– layer charge, 178, 179, 181
– octahedral cations, related to interlayer cations, 30
– octahedral sheet, cation distribution, 124
– –, composition, 51, 52, 53, 174, 175, 176
– –, –, compared to glauconite, 51, 53
– origin, 47
– polytypes, 19, 47

- structural formulas, 48, 49, 51
- structure related to composition, 183
- tetrahedral rotation, 184
- tetrahedral sheet, composition, compared to glauconite, 47

Chain-structure clays, 176, 177, 178, 183, 186, 187

Chamosite, 87, 93
-, b-axis displacement, 166
- cation distributions, 164
-, chemical analyses, 161
-, impurities in, 160
- layer charge, 164, 165
- occurrence, 160
- octahedral sheet, composition, 161, 164, 165
- origin, 165
- stacking, 165
-, structural formulas, 160
- structure, related to composition, 165
- tetrahedral sheet, composition, 160

Chemical analyses 5, 57, 69
- -, allophane, 156
- -, amesite, 164
- -, anauxite, 132
- -, attapulgite, 123
- -, beidellite, 60
- -, celadonite, 48, 49
- -, chamosite, 161
- -, chlorite, 96
- -, chlorite-montmorillonite, 115
- -, cronstedtite, 164
- -, Cr-rich trioctahedral smectite, 86
- -, Cu-rich trioctahedral smectite, 86
- -, dickite, 146
- -, glauconite, 32, 42, 44
- -, greenalite, 164
- -, halloysite, 150, 151, 152, 153
- -, hectorite, 78
- -, illite, 6, 8, 9
- -, illite-montmorillonite, 109
- -, kaolinite, 132, 134
- -, kaolinite-montmorillonite, 142
- -, montmorillonite, 58, 59, 64, 65
- -, nacrite, 145
- -, nontronite, 76
- -, palygorskite, 120
- -, rectorite, 108
- -, saponite, 80
- -, sauconite, 84
- -, sepiolite, 128
- -, serpentine-chlorite, 117
- -, stevensite, 78

- -, talc-saponite, 117

Chlorite, 4, 87–98
- cation distribution, 88
- classification, 87
- clay, 91–98
- -, dioctahedral, 94–98
- -, trioctahedral, 91–94
- diagenesis, dioctahedral, 20
- -, trioctahedral, 20
- -, Fe_2O_3 content of, 88
- hydrothermal studies, 89
- -, macroscopic, 87–91
    octahedral sheet, 173, 175
- -, composition, 88, 89
- polytypes, 89, 90, 91
- -, related to environment, 90, 91
- tetrahedral sheet, composition, 88, 89
- variations, structural, 89, 90

Chlorite clay minerals, 91–98
-, dioctahedral, 94–98
-, - chemical analyses, 96
-, - interlayers, Al and Fe, 94, 95
-, - occurrence, 91, 96, 97
-, - octahedral sheet, composition, 91
-, - origin, 94, 95, 96, 98
-, - structural formulas, 97
-, - tetrahedral sheet, composition, 97
-, - -, -, related to origin, 98
-, - X-ray data, 95, 97
-, di/trioctahedral, 95
- formation from biotite, 98
- formation from vermiculite, 105
-, trioctahedral, 91–94
-, - interlayers, Mg, 94
-, - origin, 92, 93, 94
-, tetrahedral sheet, composition, 92
-, - X-ray data, 91, 92, 93

Chlorite-montmorillonite, 114–118
- chemical analyses, 115
- formation from biotite, 98
- layer charge, 116, 117
- occurrence, 116
- stacking, 114, 116
- tetrahedral sheet, composition, 116

Chlorite-vermiculite, 91, 93, 106
- formation from biotite, 98

Chrysotile, 1, 159

Clinochlore, 87

Colloids, 170

Composition related to structure, 173, 187

Corrensite, 4, 114

Cronstedtite, chemical analysis, 164
- Fe distribution, 167

INDEX

- octahedral sheet, composition, 165
- structural formula, 164

Daphnite, 118
Diabantite, 79, 87
Diagenesis, 14, 20, 21, 37
Dickite, 145–147
- chemical analyses, 146
- composition related to origin, 146
-, impurities in, 146
- origin, 145
- structure, 145
Donbassite, 94

Endellite, 152
Evansite, 157
Expandable clay minerals, 3, 4

Ferrihalloysite, 152, 153
Fire clay, 142
Flint clay, 142

Gel, 171, 172
-, formation of glauconite, 43
- in allophane, 155, 157
Gibbsite sheet, 1
- in chlorite, 94, 95, 98
- structure related to allophane, 155
- synthesis, 170, 171, 172
- -, method, 171
Glauconite, 3, 4, 25–45
- age related to composition, 39–43
- b-axis related to composition, 30
- cation distributions, 28
- -, correlation coefficients, 36
- -, statistical data, 26
- cation exchange capacity, 38, 39
-, character of, related to lithology, 41
- charge distribution vs. K content, 34
- chemical analyses, 32
- -, average related to selected suites, 42
- -, non-marine, 44
- color, diagenesis, 37
- -, Fe^{3+}/Fe^{2+}, 37
- -, Francolite, 38
- expandable layers vs. K content, 34, 35
- interlayer cations, 27, 28, 38, 39
- -, K analyses, 42
- layer charge, 178, 179, 180
- -, compared to illite, 21
- morphology, 38
- octahedral cations, related to interlayer cations, 30

- -, frequency distribution, 34
- octahedral sheet, composition, 174, 175, 176
- -, -, compared to beidellite, 29
- -, -, compared to illite, 31
- -, -, compared to montmorillonite, 31
- -, -, compared to nontronite, 29
- origin, marine, 43, 44
- -, non-marine, 44, 45
-, oxidation of ferrous iron in, 34
- oxide distributions, 27
- oxides, correlation coefficients, 36
- -, statistical data, 26
- polytypes, 19, 25
- structural formulas, 32, 33
- -, average, 25, 27
- -, -, selected suites, 42
- tetrahedral rotation, 185
- tetrahedral sheet, composition, 28
- water content, 37
Goethite, 152
Greenalite, chemical analyses, 164
- Fe distribution, 167
- occurrence, 167
- octahedral sheet, composition, 165
- origin, 167
- structural formulas, 164

Halloysite, 149–154
- cation-exchange capacity, 152, 154
- chemical analyses, 150, 151, 152
- -, Fe-rich, 153
- composition compared to kaolinite, 150
-, dehydrated, 149
- formation from allophane, 155
-, hydrated, 149
-, impurities in, 159, 152
- morphology, 149, 152
- -, related to composition, 154
- origin, 151, 152
- structure, 149, 152, 154
- water content, 149
Hectorite, 3
- chemical analyses, 78
- octahedral sheet, composition, 79
- origin, 79
- structural formulas, 78
- tetrahedral sheet, composition, 79
Hydromica (hydromuscovite), 23

Illidromica, 18
Illite, 3, 5–23
- Al distribution, 68

- association with attapulgite, 125
- cation distributions, 15
- cation exchange capacity, 19
- cations, compared to palygorskite, 121
- –, correlation coefficients, 17
- –, chemical analyses, 6, 8, 9
- composition compared to attapulgite, 186
- diagenesis, 20
- interlayer cations, 13, 14, 16, 17, 18
- layer charge, 7, 13
- –, compared to glauconite, 31
- –, related to composition, 179, 180
- mixed-layering (montmorillonite), 13
- occurrence, 7
- octahedral sheet, composition, 174, 175
- origin, 7, 19, 20, 21
- –, related to composition, 18
- –, thermal energy, 13
- oxide distribution, 14
- oxides, correlation coefficients, 16
- polytypes, 5, 7, 13, 19, 20
- structural formulas, 10, 11
- –, average, 12
- tetrahedral rotation, 185
- –, trioctahedral, 18

Illite-chlorite, 20, 21
Illite-chlorite-montmorillonite, 114
Illite-montmorillonite, 107–114
- cation exchange capacity, 19, 110, 113
- cations, correlation coefficients, 111
- –, altered vermiculite, 106
- chemical analyses, 109
- layer charge, 74
- –, related to composition, 179, 180
- –, related to % interlayer K, 112, 113
- octahedral sheet, composition, 174, 175
- origin, 20, 107, 113, 114
- oxides, correlation coefficients, 111
- stacking, random, 107
- –, regular, 107
- –, structural formulas, 110

Imogolite, 158
Interlayer organic molecules, 3, 4
Interlayer water, 3, 4

Kaolinite, 1, 131–144
- –, alumina excess in, 131, 133
- –, amorphous material in, 143
- –, amorphous silica in, 131, 133
- cation exchange capacity, 137, 138, 141, 142, 143
- –, surface area, 143
- –, amorphous material, 143

- –, pH, 143
- –, montmorillonite, 144
- –, chemical analyses, 132, 134
- composition, compared to dickite, 146
- –, compared to halloysite, 146
- –, related to origin, 142
- –, related to particle size, 133
- Diagenesis, 20
- exchangeable cations, 144
- –, Fe, impurity, 137
- –, –, structural, 137
- formation from allophane, 155
- –, hard, 137, 138, 141
- –, –, compared to soft, 139
- –, –, correlation coefficients, 140
- –, impurities in, 131, 133, 137, 142
- origin, 141, 142
- –, residual, 141, 142
- –, sedimentary, 141, 142
- –, soft, 137, 138, 141
- –, –, compared to hard, 139
- –, –, correlation coefficients, 140
- surface area, cation exchange capacity, 143
- –, TiO_2, 133, 135
- –, water, 133, 135, 142
- synthesis, 171
- –, Ti, impurity in, 137
- –, –, structural, 137
- water, related to surface area, 133, 135, 142

Kaolinite-montmorillonite, 142
- –, chemical analysis, 142

Layer-charge distribution, 178–186
Lepidomelane, 161
Lizardite, 159
Loughlinite, 128

Metahalloysite, 149
Mixed-layer clay minerals, 4, 107–118
- layer charge, related to composition, 179, 180, 181
- –, random, definition, 4
- –, regular, definition, 4

Montmorillonite, 3, 55–75
- associated with attapulgite, 119, 125, 126
- associated with kaolinite, 137, 139, 140, 144
- cation distribution, compared to palygorskite, 121, 123
- cation exchange capacity, 69, 70
- –, compared to illite, 73
- –, chemical analyses, 58, 59
- –, amorphous material, 69

INDEX

- –, Fe-rich, 76
- –, Cheto-type, 60
- –, –, chemical analyses, 64, 65
- –, –, structural formulas, 65
- – composition compared to attapulgite, 186
- – –, compared to palygorskite, 119, 121
- – diagenesis, 20
- – exchangeable cations, 70, 71, 72
- –, related to diagenesis, 72
- –, related to environment, 72
- –, related to flake size, 72
- –, formation of clay chlorite, 94, 95
- – impurity in kaolinite, 131
- – layer charge, 179, 180
- –, manganiferous, 75
- – octahedral sheet, composition, 174, 175, 176, 177
- –, effect of Mg, 23
- – origin, 55
- – –, related to composition, 74, 75
- – structural formulas, 58, 59, 69, 73, 77
- – structure related to composition, 181
- – synthesis, 169, 170, 171
- – tetrahedral rotation, 184, 185
- –, Wyoming-type, 61
- –, –, chemical analyses, 64, 65
- –, –, structural formulas, 65

Montmorillonite-beidellite, Al distribution, 63, 66, 67
- –, affected by temperature, 67
- –, compared to illite, 66, 67, 68
- – cation distributions, 62
- – cations, correlation coefficients, 63
- – –, statistical data, 56
- – composition, effect on structure, 67
- – octahedral cations, 57
- – –, compared to illite, 57
- – oxide distributions, 61
- – correlation coefficients, 66
- – oxides, statistical data, 56
- – tetrahedral sheet, composition, 57
- – –, compared to illite, 57

Muscovite, 2, 187
- – interlayer cation, K, 67
- – octahedral sheet, 183

Nacrite, 145, 147
- – chemical analyses, 145
- – origin, 145
- – structure, 145

Nontronite, 3, 75–77
- – chemical analyses, 76

- – composition, compared to montmorillonite-beidellite, 75
- – layer charge distribution, 76, 179, 180
- – octahedral sheet, composition, 75, 174, 175, 176, 177
- – –, compared to glauconite, 29
- – origin, 77
- – –, related to glauconite, 43
- –, structural formulas, 76, 77
- – structure related to composition, 182
- – tetrahedral rotation, 184, 185
- – tetrahedral sheet, composition, 75

Octahedral sheet, 1
- –, dioctahedral, 175, 176
- –, trioctahedral, 173, 174, 175
- –, substitution in, 2, 3, 4

Palygorskite, 119–124
- – cation distribution, 121, 123
- – –, compared to 2:1 and 1:1 layers, 124
- – cation exchange capacity, 120
- – chemical analyses, 120
- – octahedral Al compared to Al in sepiolite, 129
- – octahedral sheet, composition, 119, 177
- –, structural formulas, 122
- – structure, 119
- – –, related to composition, 124
- – tetrahedral sheet, composition, 119

Pectolite, 79
Penninite, 87, 89
Phengite, 23
Phlogopite, 2
- – synthesis, 169
Polymerization, 170
Polytypes, 2
- –, celadonite, 47
- –, chlorite, 89, 90, 91
- –, glauconite, 25
- –, illite, 5, 7, 19, 20
- –, –, related to octahedral Fe, 13
- –, –, related to K_2O, 13, 14

Rectorite, 107
- – chemical analyses, 108
- – composition related to origin, 109
- –, Fourier analysis of, 109
- – interlayer cations, 109
- – origin, 109
- – structural formulas, 108
Ripidolite, 87

Saponite, 3
- cation exchange capacity, 81, 83
-, chemical analyses, 80
- octahedral cations related to interlayer cations, 30
- octahedral sheet, composition, 174, 176
- -, -, compared with trioctahedral 1:1 layers, 165
- origin, 83
- structural formulas, 81
Sauconite, chemical analyses, 84
- octahedral sheet, composition, 83, 84
- origin, 84
- structural formulas, 85
- structure related to composition, 84
- tetrahedral sheet, composition, 83
Sepiolite, 4, 127–130
- associated with attapulgite, 125
- cation distributions, 127, 128
- cation exchange capacity, 130, 178
-, chemical analyses, 128
- occurrence, 130
- octahedral cations, compared to palygorskite, 129
- octahedral sites, composition, 176, 177
- origin, 130
- structural formulas, 127, 129
- structure, 127
- -, related to compositon, 186
- synthesis, 171
- water, bound, 127, 130
- -, structural 127, 130
- -, zeolitic, 127, 130
Septechlorite, 89, 160
Sericite, 21–23
-, chemical analyses, 21
- origin, 21
- structural formulas, 22
Serpentine, 1, 2
-, Al, X-ray data on, 159, 160
- occurrence, 159
- octahedral sheet, composition, 159
- structure, related to composition, 165
- -, related to morphology, 159, 165
- tetrahedral sheet, composition, 159
- synthesis, 171
Serpentine-chlorite, chemical analyses, 117
- structural formulas, 117
Sheridanite, 87
Siderophyllite, 161
Smectite, 2, 55–86
-, dioctahedral, 55–77
-, trioctahedral, 77–86

-, -, Cr-rich, chemical analysis, 86
-, -, Cu-rich, chemical analysis, 86
Solubility, silica, 170
Stevensite, cation exchange capacity, 79
-, chemical analyses, 78
- octahedral sheet, 79
- origin, 79
- structural formulas, 78
- tetrahedral sheet, 79
Structural formulas, 72
-, amesite, 164
-, beidellite, 60
-, celadonite, 48, 49
-, chamosite, 160
-, chlorite, 97
-, cronstedtite, 164
-, glauconite, 32, 33, 42
-, hectorite, 78
-, greenalite, 164
-, illite, 10, 11, 12
-, illite-montmorillonite, 110
-, montmorillonite, 58, 59, 64, 65, 69, 73
-, nontronite, 77
-, palygorskite, 122
-, rectorite, 108
-, saponite, 81
-, sauconite, 85
-, sepiolite, 127, 129
-, sericite, 22
-, serpentine-chlorite, 117
-, stevensite, 78
-, talc-saponite, 117
-, vermiculite, 101, 103, 105
Structure of clay minerals, 1, 2, 3, 4
Structure related to composition, 173–187
-, tetrahedral rotation relative to b-axis, 184
Synthesis, amorphous material, 169, 170
-, antigorite, 169
-, bayerite, 171
-, boehmite, 171
-, brucite, 170
-, chlorite, 89
-, gibbsite, 170, 171, 172
-, kaolinite, 171
-, low-temperature, 169–172
-, method, electrolytic, 169
-, -, gibbsite, 171
-, -, montmorillonite, 171
-, montmorillonite, 169, 170, 171
-, pH effects of, 169, 170, 171, 172
-, phlogopite, 169
-, sepiolite, 171
-, serpentine, 171

INDEX

–, talc, 170

Talc, related to stevensite, 79
– synthesis, 170
Talc-saponite, chemical analyses, 117
– origin, 118
– structural formulas, 117
Taranakite, 157
Tetrahedral inversion, 183, 186
Tetrahedral rotation, 184, 185
–, in chlorite, 95
Tetrahedral sheet, 1
–, substitution in, 2, 3, 4
Thuringite, 87
Tosudite, 116
Trioctahedral 1:1 clay minerals, 159–167
– octahedral sheet, composition, 165

Vermiculite, 3, 4, 99–106
– cation exchange capacity, 99, 101
– compared to montmorillonite, 100
– defined, 99, 100
–, hydrothermal studies of, 106
– interlayer cations, 102
– interlayer water, 102
– structural formulas, 101
– octahedral sheet, composition, 100, 101
– origin, 99, 100
– –, from biotite, 99, 100, 101
– –, from phlogopite, 102
– structural formulas, 101
Vermiculite clay mineral, 102–106
– alteration to chlorite, 105
– association with chlorite, 91
– cation exchange capacity, 103
– –, related to Fe_2O_3 coating, 105
– composition related to origin, 105
– layer charge related to flake size, 105
– occurrence, 106
– origin, 102, 103
–, structural formulas, 103, 105
– tetrahedral sheet, composition, 103, 104
– –, related to origin, 104

Xylotile, 128, 129